Complexity, Global Politics, and National Security

Complexity, Global Politics, and National Security

Edited by
David S. Alberts
and
Thomas J. Czerwinski

National Defense University
Washington, D.C.

NATIONAL DEFENSE UNIVERSITY
President: Lieutenant General Ervin J. Rokke, USAF
Vice President: Ambassador William G. Walker
INSTITUTE FOR NATIONAL STRATEGIC STUDIES
Director: Dr. Hans A. Binnendijk
ADVANCED CONCEPTS, TECHNOLOGIES, AND INFORMATION STRATEGIES (ACTIS)
Director: Dr. David S. Alberts
Fort Lesley J. McNair, Washington, DC 20319-6000
Phone: (202) 685-2209 w Facsimile: (202) 685-3664

ISBN 1-57906-046-3

First printing, June 1997

Table of Contents

Part III. Complexity Theory, Strategy, and Operations

Acknowledgments

The National Defense University and RAND Corporation thank the speakers, panel members, and attendees for their participation in the Complexity, Global Politics and National Security Conference held on November 13-14, 1996. Recognition is given to the conference co-chairmen, Dr. David S. Alberts of NDU and Dr. Richard L. Kugler of RAND, and Committee members Dr. Gerry Gingrich, Dr. Jerome F. Smith, Jr., and Thomas J. Czerwinski for organizing the conference.

Invaluable support was provided by Ms. Tonya Inabinett, event coordinator, Mr. Jeffrey Beasley and YN2 Michael Chambers of the National Defense University, and Ms. Rosemaria B. Bell of Evidence Based Research, Inc.

This volume of proceedings is due to the efforts of the editors Dr. David S. Alberts and Thomas J. Czerwinski, ably assisted by Ms. Tonya Inabinett, Mr. Harry Finley, Ms. Rhonda Gross, and Mr. Juan Medrano of the National Defense University, and Ms. Lydia Candland Alexander of Evidence Based Research, Inc.

Foreword

The National Defense University was pleased to join with the RAND Corporation in sponsoring the symposium on *Complexity, Global Politics and National Security* in November 1996. I believe that these proceedings have much to offer, particularly to those of us who are associated with the profession of arms.

Gregory Treverton of RAND, in his welcoming remarks at the symposium, described the confusing times in which we live by paraphrasing Churchill's comment following an undistinguished meal, that "The pudding lacked a theme." Treverton went on to ask how without a theme, do we apprehend, how do we understand this world?

In trying to answer that question, I think it is fair to say that the intellectual response to the end of the Cold War has tended by-and-large to focus on what is called the Revolution in Military Affairs. This is driven by advances in technology, primarily information technology. The discussions about the Revolution in Military affairs are interesting and important. However, to my taste, what emerges is a "pudding without a theme."

We have given less attention to what our colleagues in the arenas of physics, biology and other New Sciences have to say. They suggest that neither technology nor the Newtonian principles of linearity are sufficient to deal with the increasingly complex world in which we find ourselves. Complexity theory contends that there are underlying simplicities, or patterns, if we but look for them. These provide us with insights, if not predictions and solutions. Such an effort, if successful, promises to help us find the theme in the pudding.

I believe that we have done some things and made some progress, thanks in particular to the U.S. Marine Corps, in the application of nonlinear principles to the battlefield and operational art. Hopefully that progress will continue. But we need to devote our focus and concerns about the impact of nonlinearity on the arenas of strategy and international relations, as well. These proceedings help to move us in that direction.

As I urged the symposium's audience, I now urge the proceedings readers, "Kick off your mental shoes, and let your minds stray out of the boxes into which we normally find ourselves." See if among these papers there is a theme in the pudding.

Ervin J. Rokke
Lieutenant General, US Air Force
President, National Defense University

Preface

The emergence of Complexity theory on the national security scene should come as no surprise. In fact, it is rather late arriving compared to such fields as corporate management, economics and markets, and ecology, among others. This can be attributed to a belated recognition of its potential by both National Security practitioners and Complexity theorists.

Complexity theory can be viewed as the native form for investigating the properties and behavior of the dynamics of nonlinear systems. This stands in contrast to the non-native modes invented by the linear domain to probe the largely nonlinear world around us—calculus, statistics, rounding and rules of thumb.

By linear systems, we mean the arrangement of nature—life and its complications—to be one where outputs are proportional to inputs; where the whole is equal to the sum of its parts, and where cause and effect are observable. It is an environment where prediction is facilitated by careful planning; success is pursued by detailed monitoring and control; and a premium is placed upon reductionism, rewarding those who excel in reductionist processes. Reductionist analysis consists of taking large, complex problems and reducing them to manageable chunks.

By nonlinear systems, we mean the arrangement of nature—life and its complications, such as warfare—in which inputs and outputs are not proportional; where the whole is not quantitatively equal to its parts, or even, qualitatively, recognizable in its constituent components; and where cause and effect are not evident. It is an environment where phenomena are unpredictable, but within bounds, self-organizing; where unpredictability frustrates conventional planning, where solution as self-organization defeats control; and where the "bounds" are the actionable variable, requiring new ways of thinking and acting.

The inquiry into the nature of nonlinearity, and the rise of Complexity theory has of necessity paralleled the development of the computer. Nonlinearity is extremely difficult to work with unless aided by the computer. Nonlinear equations were referred to as the "Twilight Zone" of mathematics. Beginning in the early 1960s, efforts to modify the weather indicated the severe limits to predictability in nonlinear environments, such as weather, itself. The self-organizing nature of nonlinearity, and the attributes of Chaos theory were well advanced by 1987, with the publication of James Gleick's best-selling popularization *Chaos: Making a New Science.* In the mid-1980s, the Santa Fe Institute was organized to further the inquiry into complex adaptive systems. By 1992, Complexity theory also qualified for publication in the popular press with Mitchell Waldrop's Complexity: *The Emerging Science at the Edge of Order and Chaos*, and Steven Lewin's *Complexity: Life at the Edge of Chaos.* Nonlinearity was now in the public domain and universally accessible.

A number of modern U.S. defense thinkers, in retrospect, can be considered to be nonlinearists. Prominent among these are J.C. Wylie and the prolific, but unpublished, John Boyd of OODA loop fame. However, in the context of the time and vocabulary, this realization could only be implicit. An explicit articulation only began to emerge in the early 1990s. Two of the earliest pioneers are authors in this volume. Both wrote seminal papers, the significance of which was largely unrecognized when they first appeared. In late 1992, Alan Beyerchen's "Clausewitz, Nonlinearity, and the Unpredictability of War," was published in *International Security*, and Steven Mann's "Chaos Theory and Strategic Thought" appeared in *Parameters*. The former work is a profound reinterpretation of Clausewitz's *On War*, persuasively placing the work, and Clausewitz, himself, in a nonlinear framework. Mann, a Foreign Service officer, used self-organizing criticality, a concept associated with the Santa Fe Institute, to describe the dynamics of international relations and its implications for strategy.

These initial intellectual contributions were followed by important advances, each the individual efforts of talented Air Force officers. These included investigations into defense applications of Chaos theory (David Nicholls, et al.,1994, and Glenn E. James,1995.) Paralleling these efforts were those in Complexity theory applied to the determination of centers of gravity (Pat A. Pentland, 1993), and especially a robust and detailed methodology for identifying target sets (Steven M. Rinaldi, 1995). As a result, the confidence factor rose appreciably, as the body of defense-related literature began to assume the qualitative and quantitative dimensions for a discipline, or a contending body of thought. Primarily at the operational and tactical levels of war, nonlinear concepts were moving beyond the notional, to formulation and application.

A major breakthrough came in 1994, when the U.S. Marine Corps adopted nonlinear dynamics, and the ideas of Complexity theory, realizing that they provided an underlying basis for the Marine doctrine of maneuver warfare embodied in the capstone manual *Warfighting*. In a sense, science came to abet the school of hard-knocks and experience. This has triggered a host of ongoing exciting innovations and initiatives, notably the 1996 publication of MCDP 6-*Command and Control*, which explicitly rests on Complexity theory concepts. Nevertheless, the application of Complexity still lagged in the policy and strategic domains of the national security arena.

It is against this background that the symposium was held. The charge given by the President of the National Defense University and RAND leadership was threefold: (1) Capitalize on the momentum described above, and push the envelope; (2) Emphasize the nonlinearity of international affairs, and the policy and strategic dimensions of national defense with the implications for Complexity theory; and (3) Get the best talent available in academe.

These papers were first delivered at the two-day symposium held at the National Defense University in November 1996. In addition to the contributors to this volume, other speakers included Richard L. Kugler, Paul K. Davis, and Carl H. Builder of RAND. Importantly, VADM Arthur K. Cebrowski, USN, LTG John E. Miller, USA, LtGen Ervin J. Rokke, USAF, and LtGen Paul K. Van Riper, USMC were in constant attendance forming an Operations Perspectives Panel. Their invaluable participation throughout helped to shape the symposium, by honing its perspective for that of the Warrior.

David S. Alberts
Director, Advanced Concepts, Technologies and Information Strategies

Thomas J. Czerwinski
Professor, School of Information Warfare and Strategy

PART ONE

SETTING THE SCENE

The Simple and the Complex

Murray Gell-Mann

It is a pleasure, as well as an honor, to give the opening talk at this conference on Complexity, Global Politics, and National Security. I am glad to be paying my first visit to the National Defense University. As to the other sponsoring institution, I am no stranger to it. In fact, it is just forty years since I first became a RAND consultant. Now both organizations have become interested in such concepts as chaos and complexity, and I am delighted to have the opportunity to discuss them here.

At the Santa Fe Institute, which I helped to found and where I now work, we devote ourselves to studying, from many different points of view, the transdisciplinary subject that includes the meanings of simplicity and complexity, the ways in which complexity arises from fundamental simplicity, and the behavior of complex adaptive systems, along with the features that distinguish them from non-adaptive systems.

My name for that subject is *plectics*, derived from the Greek word *plektós* for "twisted" or "braided," cognate with the principal root of Latin *complexus*, originally "braided together," from which the English word *complexity* is derived. The word *plektós* is also related, more distantly, to the principal root of Latin *simplex*, originally "once folded," which gave rise to the English word *simplicity*. The name *plectics* thus reflects the fact that we are dealing with both simplicity and complexity.

I believe my task this morning is to throw some light on plectics and to indicate briefly how it may be connected with questions of national and global security, especially when the term "security" is interpreted rather broadly. We can begin with questions such as these:

- •What do we usually mean by complexity?

- •What is chaos?

- •What is a complex adaptive system?

Why is there a tendency for more and more complex entities to appear as time goes on?

It would take a number of quantities, differently defined, to cover all our intuitive notions of the meaning of complexity and of its opposite, simplicity. Also, each quantity would be somewhat context-dependent. In other words, complexity, however defined, is not entirely an intrinsic property of the entity described; it also depends to some extent on who or what is doing the describing.

Let us start with a rather naïvely defined quantity, which I call "crude complexity"—the length of the shortest message describing the entity. First of all, we would have to exclude pointing at the entity or calling it by a special name; something that is obviously very complex could be given a short nickname like Heinz or Zbig, but giving it that name would not make it simple. Next, we must understand that crude complexity will depend on the level of detail at which the entity is being described, what we call in physics the coarse graining. Also, the language employed will affect the minimum length of the description. That minimum length will depend, too, on the knowledge and understanding of the world that is assumed: the description of a rhinoceros can be abbreviated if it is already known what a mammal is.

Having listed these various kinds of context dependence, we can concentrate on the main feature of crude complexity, that it refers to length of the shortest message. In my book, *The Quark and the Jaguar*, I tell the story of the elementary school teacher who assigned to her class a three hundred-word essay, to be written over the weekend, on any topic. One pupil did what I used to do as a child—he spent the weekend poking around outdoors and then scribbled something hastily on Monday morning. Here is what he wrote:

> "Yesterday the neighbors had a fire in their kitchen and I leaned out of the window and yelled 'Fire! Fire! Fire! Fire!...'" If he had not had to comply with the three hundred

word requirement, he could have written instead "...I leaned out of the window and yelled 'Fire!' 282 times."

It is this notion of *compression* that is crucial.

Now in place of crude complexity we can consider a more technically defined quantity, algorithmic information content. An entity is described at a given level of detail, in a given language, assuming a given knowledge and understanding of the world, and the description is reduced by coding in some standard manner to a string of bits (zeroes and ones). We then consider all programs that will cause a standard universal computer to print out that string of bits and then stop computing. The length of the shortest such program is called the algorithmic information content (AIC). This is a well-known quantity introduced over thirty years ago by the famous Russian mathematician Kolmogorov and by two Americans, Gregory Chaitin and Ray Solomonoff, all working independently. We see, by the way, that it involves some additional context dependence through the choice of the coding procedure and of the universal computer. Because of the context dependence, AIC is most useful for comparison between two strings, at least one of which has a large value of it.

A string consisting of the first two million bits of pi has a low AIC because it is highly compressible: the shortest program just has to give a prescription for calculating pi and ask that the string be cut off after two million entries. But many long strings of bits are incompressible. For those strings, the shortest program is one that lists the whole string and tells the machine to print it out and then halt. Thus, for a given length of string, an incompressible one has the largest possible AIC. Such a string is called a "random" one, and accordingly the quantity AIC is sometimes called algorithmic randomness.

We can now see why AIC does not correspond very well to what we usually mean by complexity. Compare a play by Shakespeare with the typical product, of equal length, of the proverbial ape at the typewriter, who types every letter with equal probability. The AIC, or algorithmic randomness, of the latter is much greater than that of the former. But it is absurd to say that the ape has produced something more complex than the work of Shakespeare. Randomness is not what we mean by complexity.

Instead, let us define what I call effective complexity, the AIC of the *regularities* of an entity, as opposed to its incidental features. A random (incompressible) bit string has no regularities (except its length) and very little effective complexity. Likewise something extremely regular, such as a bit string consisting entirely of ones, will also have very little effective complexity, because its regularities can be described very briefly. To achieve high effective complexity, an entity must have intermediate AIC and obey a set of rules requiring a long description. But that is just what we mean when we say that the grammar of a certain language is complex, or that a certain conglomerate corporation is a complex organization, or that the plot of a novel is very complex—we mean that the description of the regularities takes a long time.

The famous computer scientist, psychologist, and economist Herbert Simon used to call attention to the path of an ant, which has a high AIC and appears complex at first sight. But when we realize that the ant is following a rather simple program, into which are fed the incidental features of the landscape and the pheromone trails laid down by the other ants for the transport of food, we understand that the path is fundamentally not very complex. Herb says, "I got a lot of mileage out of that ant." And now it is helping me to illustrate the difference between crude and effective complexity.

There can be no finite procedure for finding all the regularities of an entity. We may ask, then, what kinds of things engage in identifying sets of regularities. The answer is: complex adaptive systems, including all living organisms on Earth.

A complex adaptive system receives a stream of data about itself and its surroundings. In that stream, it identifies particular regularities and compresses them into a concise "schema," one of many possible ones related by mutation or substitution. In the presence of further data from the stream, the schema can supply descriptions of certain aspects of the real world, predictions of events that are to happen in the real world, and prescriptions for behavior of the complex adaptive system in the real world. In all these cases, there are real world consequences: the descriptions can turn out to be more accurate or less accurate, the predictions can turn out to be more reliable or less reliable, and the prescriptions for behavior can turn out to lead to favorable or unfavorable outcomes. All these consequences then feed back to exert "selection pressures" on the competition among various schemata, so that there is a strong tendency for more successful schemata to survive and for less successful ones to disappear or at least to be demoted in some sense.

Take the human scientific enterprise as an example. The schemata are theories. A theory in science compresses into a brief law (say a set of equations) the regularities in a vast, even indefinitely large body of data. Maxwell's equations, for instance, yield the electric and magnetic fields in any region of the universe if the special circumstances there—electric charges and currents and boundary conditions—are specified. (We see how the schema plus additional information from the data stream leads to a description or prediction.)

In biological evolution, the schemata are genotypes. The genotype, together with all the additional information supplied by the process of development—for higher animals, from the sperm and egg to the adult organism—determines the character, the "phenotype," of the individual adult. Survival to adulthood of that individual, sexual selection, and success or failure in producing surviving progeny all exert selection pressures on the competition of genotypes, since they affect the transmission to future generations of genotypes resembling that of the individual in question.

In the case of societal evolution, the schemata consist of laws, customs, myths, traditions, and so forth. The pieces of such a schema are often called "memes," a term introduced by Richard Dawkins by analogy with genes in the case of biological evolution.

For a business firm, strategies and practices form the schemata. In the presence of day-to-day events, a schema affects the success of the firm, as measured by return to the stockholders in the form of dividends and share prices. The results feed back to affect whether the schema is retained or a different one substituted (often under a new CEO).

A complex adaptive system (CAS) may be an integral part of another CAS, or it may be a loose aggregation of complex adaptive systems, forming a composite CAS. Thus a CAS has a tendency to give rise to others.

On Earth, all complex adaptive systems seem to have some connection with life. To begin with, there was the set of prebiotic chemical reactions that gave rise to the earliest life. Then the process of biological evolution, as we have indicated, is an example of a CAS. Likewise each living organism is a CAS. In a mammal, such as a human being, the immune system is a complex adaptive system too. Its operation is something like that of biological evolution, but on a much faster time scale. (If it took hundreds of thousands of years for us to develop antibodies to invading microbes, we would be in serious trouble.) The process of learning and thinking in a human individual is also a complex adaptive system. In fact, the term "schema" is taken from psychology, where it refers to a pattern used by the mind to grasp an aspect of reality. Aggregations of human beings can also be complex adaptive systems, as we have seen: societies, business firms, the scientific enterprise, and so forth.

Nowadays, we have computer-based complex adaptive systems, such as "neural nets" and "genetic algorithms." While they may sometimes involve new, dedicated hardware, they are usually implemented on conventional hardware with special software. Their only connection with life is that they were developed by human beings. Once they are put into operation, they can, for example, invent new strategies for winning at games that no human being has ever discovered.

Science fiction writers and others may speculate that in the distant future a new kind of complex adaptive system might be created, a truly composite human being, by wiring together the brains of a number of people. They would communicate not through language, which Voltaire is supposed to have said is used by men to conceal their thoughts, but through sharing all their mental processes. My friend Shirley Hufstedler says she would not recommend this procedure to couples about to be married.

The behavior of a complex adaptive system, with its variable schemata undergoing evolution through selection pressures from the real world, may be contrasted with "simple" or "direct" adaptation, which does not involve a variable schema, but utilizes instead a fixed pattern of response to external changes. A good example of direct adaptation is the operation of a thermostat, which simply turns on the heat when the temperature rises above a fixed value and turns it off when the temperature falls below the same value.

In the study of a human organization, such as a tribal society or a business firm, one may encounter at least three different levels of adaptation, on three different time scales.

1) On a short time scale, we may see a prevailing schema prescribing that the organization react to particular external changes in specified ways; as long as that schema is fixed, we are dealing with direct adaptation.

2) On a longer time scale, the real world consequences of a prevailing schema (in the presence of events that occur) exert selection pressures on the competition of schemata and may result in the replacement of one schema by another.

3) On a still longer time scale, we may witness the disappearance of some organizations and the survival of others, in a Darwinian process. The evolution of schemata was inadequate in the former cases, but adequate in the latter cases, to cope with the changes in circumstances.

It is worth making the elementary point about the existence of these levels of adaptation because they are often confused with one another. As an example of the three levels, we might consider a prehistoric society in the U.S. Southwest that had the custom (1) of moving to higher elevations in times of unusual heat and drought. In the event of failure of this pattern, the society might try alternative schemata (2) such as planting different crops or constructing an irrigation system using water from far away. In the event of failure of all the schemata that are tried, the society may disappear (3), say with some members dying and the rest dispersed among other societies that survive. We see that in many cases failure to cope can be viewed in terms of the evolutionary process not being able to keep pace with change.

Individual human beings in a large organization or society must be treated by the historical sciences as playing a dual role. To some extent they can be regarded statistically, as units in a system. But in many cases a particular person must be treated as an individual, with a personal influence on history. Those historians who tolerate discussion of contingent history (meaning counterfactual histories in addition to the history we experience) have long argued about the extent to which broad historical forces eventually "heal" many of the changes caused by individual achievements—including negative ones, such as assassinations.

A history of the U.S. Constitutional Convention of 1787 may make much of the conflicting interests of small states and large states, slave states and free states, debtors and creditors, agricultural and urban populations, and so forth. But the compromises invented by particular individuals and the role that such individuals played in the eventual ratification of the Constitution would also be stressed. The outcome could have been different if certain particular people had died in an epidemic just before the Convention, even though the big issues would have been the same.

How do we think about alternative histories? Is the notion of alternative histories a fundamental concept?

The fundamental laws of nature are:
    (1) the dynamical law of the elementary par
    ticles—the building blocks of all matter—
    along with their interactions and
    (2) the initial condition of the universe near
    the beginning of its expansion some ten
    billion years ago.

Theoretical physicists seem to be approaching a real understanding of the first of these laws, as well as gaining some inklings about the second one. It may well be that both are rather simple and knowable, but even if we learn what they are, that would not permit us, even in principle, to calculate the history of the universe. The reason is that fundamental theory is probabilistic in character (contrary to what one might have thought a century ago). The theory, even if perfectly known, predicts not one history of the universe but probabilities for a huge array of alternative histories, which we may conceive as forming a branching tree, with probabilities at all the branchings. In a short story by the great Argentine writer Jorge Luis Borges, a character creates a model of these branching histories in the form of a garden of forking paths.

The particular history we experience is co-determined, then, by the fundamental laws and by an inconceivably long sequence of chance events, each of which could turn out in various ways. This fundamental indeterminacy is exacerbated for any observer—or set of observers, such as the human race—by ignorance of the outcomes of most of the chance events that have already occurred, since only a very limited set of observations is available. Any observer sees only an extremely coarse-grained history.

The phenomenon of *chaos* in certain nonlinear systems is a very sensitive dependence of the outcome of a process on tiny details of what happened earlier. When chaos is present, it still further amplifies the indeterminacy we have been discussing.

Last year, at the wonderful science museum in Barcelona, I saw an exhibit that beautifully illustrated chaos. A nonlinear version of a pendulum was set up so that the visitor could hold the bob and start it out in a chosen position and with a chosen velocity. One could then watch the subsequent motion, which was also recorded with a pen on a sheet of paper. The visitor was then invited to seize the bob again and try to imitate exactly the previous initial position and velocity. No matter how carefully that was done, the subsequent motion was quite different from what it was the first time. Comparing the records on paper confirmed the difference in a striking way.

I asked the museum director what the two men were doing who were standing in a corner watching us. He replied, "Oh, those are two Dutchmen waiting to take away the chaos." Apparently, the exhibit was about to be dismantled and taken to Amsterdam. But I have wondered ever since whether the services of those two Dutchmen would not be in great demand across the globe, by organizations that wanted their chaos taken away.

Once we view alternative histories as forming a branching tree, with the history we experience co-determined by the fundamental laws and a huge number of accidents, we can ponder the accidents that gave rise to the people assembled in this room. A fluctuation many billions of years ago produced our galaxy, and it was followed by the accidents that contributed to the formation of the solar system, including the planet Earth. Then there were the accidents that led to the appearance of the first life on this planet, and the very many additional accidents that, along with natural selection, have shaped the course of biological evolution, including the characteristics of our own subspecies, which we call, somewhat optimistically, Homo sapiens. Finally we may consider the accidents of genetics and sexual selection that helped to produce the genotypes of all the individuals here, and the accidents in the womb, in childhood, and since that have helped to make us what we are today.

Now most accidents in the history of the universe don't make much difference to the coarse-grained histories with which we are concerned. If two oxygen molecules in the atmosphere collide and then go off in one pair of directions or another, it usually makes no difference. But the fluctuation that produced our galaxy, while it too may have been insignificant on a cosmic scale, was of enormous importance to anything *in* our galaxy. Some of us call such a chance event a "frozen accident."

I like to quote an example from human history. When Arthur, the elder brother of King Henry VIII of England, died—no doubt of some quantum fluctuation—early in the sixteenth century, Henry replaced Arthur as heir to the throne and as the husband of Catherine of Aragón. That accident influenced the way the Church of England separated from the Roman Catholic Church (although the separation itself might have occurred anyway) and changed the history of the English and then the British monarchy, all the way down to the antics of Charles and Diana.

It is the frozen accidents, along with the fundamental laws, that give rise to regularities and thus to effective complexity. Since the fundamental laws are believed to be simple, it is mainly the frozen accidents that are responsible for effective complexity. We can relate that fact to the tendency for more and more complex entities to appear as time goes on.

Of course there is no rule that everything must increase in complexity. Any individual entity may increase or decrease in effective complexity or stay the same. When an organism dies or a civilization dies out, it suffers a dramatic decrease in complexity. But the envelope of effective complexity keeps getting pushed out, as more and more complex things arise.

The reason is that as time goes on frozen accidents keep accumulating, and so more and more effective complexity is possible. That is true even for non-adaptive evolution, as in galaxies, stars, planets, rocks, and so forth. It is well-known to be true of biological evolution, where in some cases higher effective complexity probably confers an advantage. And we see all around us the appearance of more and more complex regulations, instruments, computer software packages, and so forth, even though in many cases certain things are simplified.

The tendency of more and more complex forms to appear in no way contradicts the famous second law of thermodynamics, which states that for a closed (isolated) system, the average disorder ("entropy") keeps increasing. There is nothing in the second law to prevent local order from increasing, through various mechanisms of self-organization, at the expense of greater disorder elsewhere. (One simple and widespread mechanism of self-organization on a cosmic scale is provided by gravitation, which has caused material to condense into the familiar structures with which astronomy is concerned, including our own planet.)

Here on Earth, once it was formed, systems of increasing complexity have arisen as a consequence of the physical evolution of the planet over some four and half billion years, biological evolution over four billion years or so, and, over a very short period on a geological time scale, human cultural evolution.

The process has gone so far that we human beings are now confronted with immensely complex ecological and social problems, and we are in urgent need of better ways of dealing with them. When we attempt to tackle such difficult problems, we naturally tend to break them up into more manageable pieces. That is a useful practice, but it has serious limitations.

When dealing with any nonlinear system, especially a complex one, it is not sufficient to think of the system in terms of parts or aspects identified in advance, then to analyze those parts or aspects separately, and finally to combine those analyses in an attempt to describe the entire system. Such an approach is not, by itself, a successful way to understand the behavior of the system. In this sense there is truth in the old adage that the whole is more than the sum of its parts.

Unfortunately, in a great many places in our society, including academia and most bureaucracies, prestige accrues principally to those who study carefully some aspect of a problem, while discussion of the big picture is relegated to cocktail parties. It is of crucial importance that we learn to supplement those specialized studies with what I call a crude look at the whole.

Now the chief of an organization, say a head of government or a CEO, has to behave as if he or she is taking into account all the aspects of a situation, including the interactions among them, which are often strong. It is not so easy, however, for the chief to take a crude look at the whole if everyone else in the organization is concerned only with a partial view.

Even if some people are assigned to look at the big picture, it doesn't always work out. A few months ago, the CEO of a gigantic corporation told me that he had a strategic planning staff to help him think about the future of the business, but that the members of that staff suffered from three defects:

1) They seemed largely disconnected from the rest of the company.
2) No one could understand what they said.
3) Everyone else seemed to hate them.

Despite such experiences, it is vitally important that we supplement our specialized studies with serious attempts to take a crude look at the whole.

At this conference, issues of global politics and security will be addressed, including ones specifically concerned with the security of the United States. But security narrowly defined depends in very important ways on security in the broadest sense. Some politicians deeply concerned about military strength appear to resent the idea of diluting that concern by emphasizing a broader conception of security, but many thinkers in the armed services themselves recognize that military security is deeply intertwined with all the other major global issues.

I like to discuss those issues under the rubric of *sustainability*, one of today's favorite catchwords. It is rarely defined in a careful or consistent way, so perhaps I can be forgiven for attaching to it my own set of meanings. Broadly conceived, sustainability refers to quality that is not purchased mainly at the expense of the future—quality of human life and of the environment. But I use the term in a much more inclusive way than most people: sustainability is not restricted to environmental, demographic, and economic matters, but refers also to political, military, diplomatic, social, and institutional or governance issues—and ultimately sustainability depends on ideological issues and lifestyle choices. As used here, sustainability refers as much to sustainable peace, sustainable preparedness for possible conflict, sustainable global security arrangements, sustainable democracy and human rights, and sustainable communities and institutions as it does to sustainable population, economic activity, and ecological integrity.

All of these are closely interlinked, and security in the narrow sense is a critical part of the mix. In the presence of destructive war, it is hardly possible to protect nature very effectively or to keep some important human social ties from dissolving. Conversely, if resources are abused and human population is rapidly growing, or if communities lose their cohesion, conflicts are more likely to occur. If huge and conspicuous inequalities are present, people will be reluctant to restrain quantitative economic growth in favor of qualitative growth as would be required to achieve a measure of economic and environmental sustainability. At the same time, great inequalities may provide the excuse for demagogues to exploit or revive ethnic or class hatreds and provoke deadly conflict. And so forth.

In my book, *The Quark and the Jaguar*, I suggest that studies be undertaken of possible paths toward sustainability (in this very general sense) during the course of the next century, in the spirit of taking a crude look at the whole. I employ a modified version of a schema introduced by my friend James Gustave Speth, then president of the World Resources Institute and now head of the United Nations Development Program. The schema involves a set of interlinked transitions that have to occur if the world is to switch over from present trends toward a more sustainable situation:

1) The demographic transition to a roughly stable human population, worldwide and in each broad region. Without that, talk of sustainability seems almost pointless.

2) The technological transition to methods of supplying human needs and satisfying human desires with much lower environmental impact per person, for a given level of conventional prosperity.

3) The economic transition to a situation where growth in quality gradually replaces growth in quantity, while extreme poverty, which cries out for quantitative growth, is alleviated. (Analysts, by the way, are now beginning to use realistic measures of well-being that

depart radically from narrow economic measures by including mental and physical health, education, and so forth.) The economic transition has to involve what economists call the internalization of externalities: prices must come much closer to reflecting true costs, including damage to the future.

4) The social transition to a society with less inequality, which, as remarked before, should make the decline of quantitative growth more acceptable. (For example, fuel taxes necessary for conservation adversely affect the poor who require transport to work, but the impact of such taxes can be reduced by giving a subsidy to the working poor—such as a negative income tax—that is not tied to fuel consumption.) The social transition includes a successful struggle against large-scale corruption, which can vitiate attempts to regulate any activity through law.

5) The institutional transition to more effective means of coping with conflict and with the management of the biosphere and human activities in it. We are now in an era of simultaneous globalization and fragmentation, in which the relevance of national governments is declining somewhat, even though the power to take action is still concentrated largely at that level. Most of our problems involving security—whether in the narrow or the broad sense—have global implications and require transnational institutions for their solution. We already have a wide variety of such institutions, formal and informal, and many of them are gradually gaining in effectiveness. But they need to become far more effective. Meanwhile, local and national

institutions need to become more responsive and, in many places, much less corrupt. Such changes require the development of a strong sense of community and responsibility at many levels, but in a climate of political and economic freedom. How to achieve the necessary balance between cooperation and competition is the most difficult problem at every level.

6) The informational transition. Coping on local, national, and transnational levels with technological advances, environmental and demographic issues, social and economic problems, and questions of international security, as well as the strong interactions among all of them, requires a transition in the acquisition and dissemination of knowledge and understanding. Only if there is a higher degree of comprehension, among ordinary people as well as elite groups, of the complex issues facing humanity is there any hope of achieving sustainable quality. But most of the discussions of the new digital society concentrate on the dissemination and storage of information, much of it misinformation or badly organized information, rather than on the difficult and still poorly rewarded work of converting that so-called information into knowledge and understanding. And here again we encounter the pervasive need for a crude look at the whole.

7) The ideological transition to a world view that combines local, national, and regional loyalties with a "planetary consciousness," a sense of solidarity with all human beings and, to some extent, all living things. Only by acknowledging the interdependence of all

people and, indeed, of all life can we hope to broaden our individual outlooks so that they reach out in time and space to embrace the vital long-term issues and worldwide problems along with immediate concerns close to home. This transition may seem even more Utopian than some of the others, but if we are to manage conflict that is based on destructive particularism, it is essential that groups of people that have traditionally opposed one another acknowledge their common humanity. Such a progressive extension of the concept of "us" has, after all, been a theme in human history from time immemorial. One dramatic manifestation is the greatly diminished likelihood over the last fifty years of armed conflict in Western Europe. Another is, of course, the radical transformation of relationships that is often called "The End of the Cold War." The recent damping-down of long-standing civil wars in a number of countries is also rather impressive.

Our tendency is to study separately the various aspects of human civilization that correspond to the different transitions. Moreover, in our individual political activities we tend to pick out just one or a few of these aspects. Some of us may belong to organizations favoring a strong defense or arms control or both, others to the United Nations Association of the United States, others to ZPG or the Population Council, some to organizations plumping for more assistance to developing countries or to ones working for more generous treatment of the poor in our own country, some to organizations promoting democracy and human rights, some to environmental organizations. But the issues dear to these various organizations are all tightly interlinked, and a portion of our activity needs to be devoted to examining the whole question of the approach to sustainability in all these different spheres.

It is reasonable to ask why a set of transitions to greater sustainability should be envisaged as a possibility during the coming century. The answer is that we are living in a very special time. Historians tend to be skeptical of most claims that a particular age is special, since such claims have been made so often. But this turn of the millennium really is special, not because of our arbitrary way of reckoning time but because of two related circumstances:

a) The changes that we humans produce in the biosphere, changes that were often remarkably destructive even in the distant past when our numbers were few, are now of order one. We have become capable of wiping out a very large fraction of humanity—and of living things generally—if a full-scale world war should break out. Even if it does not, we are still affecting the composition of the atmosphere, water resources, vegetation, and animal life in profound ways around the planet. While such effects of human activities have been surprisingly great in the past, they were not global in scope as they are now.

b) The graph of human population against time has the highest rate of increase ever, and that rate of increase is just beginning to decline. In other words, the curve is near what is called a "point of inflection."

For centuries, even millennia, world population was, to a fair approximation, inversely proportional to 2025 minus the year. (That is a solution of the equation in which the rate of change of a variable is proportional to its square.) Only during the last thirty years or so has the total number of human beings been deviating significantly from this formula, which would have had it becoming infinite a generation from now! The demographic transition thus appears to be under way at last. It is generally expected that world population will level off during the coming century at something like twice its present value, but decisions and events in the near future can affect the final figure by billions either way. That is especially significant in regions such as Africa, where present trends indicate a huge population increase very difficult to support and likely to contribute to severe environmental degradation. In general, the coming century, the century of inflection points in a number of crucial variables, seems to be the time when the human race might still accomplish the transitions to greater sustainability without going through disaster.

It is essential, in my opinion, to make some effort to search out in advance what kinds of paths might lead humanity to a reasonably sustainable and desirable world during the coming decades. And while the study of the many different subjects involved is being pursued by the appropriate specialists, we need to supplement that study with interdisciplinary investigations of the strong interdependence of all the principal facets of the world situation. In short, we need a crude look at the whole, treating global security and global politics as parts of a very general set of questions about the future.

America in the World Today

Zbigniew Brzezinski

In my invitation to appear here this evening, it clearly states that, "You are not expected to deliver a lecture on Complexity theory. We merely ask you to present your views." I take it then, that this was an injunction to be simple—to provide some relief from the Complexity theory. It is in that spirit that I will share my thoughts with you regarding America's involvement in the world today. As I said, it will be simple. I will start with a simple invocation, using the basic metaphor that was the theme of the elections four years ago, "It's the economy, stupid." My invocation is, "It's leadership, stupid." That is to say that the United States has no choice—literally has no choice—but to exercise leadership in world affairs. It is not a question of whether we want to or not, it is a question that we must—literally, must. I want to stress that point because in recent times there has been a significant change in our psychological posture, as a nation.

NOTE: This published version is an edited transcript of remarks delivered without a formal text.

We have been sometimes accused, and we have in-
dicted ourselves, for having blindly followed the precept
that, "Just don't stand there, do something." We have
replaced that with a doctrine of "Don't do anything.
Just stand there and deliberate about the exit." That
is our doctrine, and I submit to you that the concept of
the "exit strategy" epitomizes a posture which is incompatible with the dilemmas that we confront on the
world scene, and the kind of leadership that we have
to find.

Let me suggest that the leadership is particularly
needed regarding six large issues, none of which can
be approached with an exit strategy. In fact, the very
concept of an exit strategy is irrelevant to the effective
addressing of these issues. The first is will a larger
and a more secure Europe emerge? The second is
will Russia become a status quo power? The third is
will the Persian Gulf and the Middle Eastern region
become more stable? The fourth is will the Far East
adjust to the very nature of the power shift that is now
under way? The fifth is will we manage effectively
nuclear proliferation? The sixth is will large-scale social collapse be avoided in some critical parts of the
world?

These are, broadly speaking, the six major issues that
we confront on the world scene. Each of these six
issues requires American engagement, and in every
one of them American leadership is necessary. Regarding none of them can we begin with, "What is the
exit strategy?"

Let us start with the first issue, "Will a larger and more secure Europe emerge?" That is clearly one of the central issues that confront us now, in the wake of the end of the Cold War. That has two basic dimensions to it. One pertains to the extension of Europe, and the other to the implications of the unification of Europe. On the extension of Europe, I believe we have made a more or less basic commitment. The President, in the course of his election campaign, made a statement which was widely publicized by the White House. It was quite explicit that it is the policy of the United States to seek the extension of the trans-Atlantic alliance by embracing several new members from Central Europe, with their membership to be attained, as an American objective, by April 1999.

I believe this to be a legitimate commitment. I do not accept the idea that this was merely an election ploy. To suggest that would be demeaning, and inaccurate. It reflects a decision reached after much deliberation, and from my point of view, too much hesitation over too long a time. But, a conscious choice nevertheless. It is my sense that the President is genuinely committed to this objective. This is the inference I gather from the very explicit character of the statement, but also in conversations with him. It is my view that his immediate advisors partake of the same commitment, some even earlier than he. I have the feeling that the Secretary of Defense is committed to that objective, and, in fact, the machinery of the Defense Department is in full gear working towards that end. I have the strong impression that the National Security

Advisor is very much committed to that idea, and has been for some time. I know that the Secretary of State, and his deputy, are in favor of the idea, about which the deputy has lately given some very significant and strong speeches. So my view is this is now our national objective.

However, it will only be attained if the United States exercises leadership. Without American leadership, we will not get there by April 1999. We will not get there by any date, at all.

Only if American leadership is firm, creative, persistent, and decisive will we make progress, not only in obtaining an alliance commitment to the objective, but in pushing forward the negotiations, in obtaining the ratification of an agreement by our own Congress, but also by the parliaments of the fifteen other members, and consummate the process by the date's end. Without strong American leadership, and also German, we will not get there. German support is very important, but German support is basically there. In fact, if American leadership is not exercised, it will be a major defeat, and will be perceived as such abroad. The German Minister of Defense told me that if we fail in pushing this purpose forward, it would have a very negative impact on our credibility.

The process of moving forward on the enlargement of Europe will engage us automatically in the equally difficult and challenging process pertaining to the unification of Europe, and that objective is just as important. On that issue, we may encounter growing difficulties in two different ways. First of all, certain European states, particularly France, will insist that any extension of NATO be accompanied simultaneously by the reform of NATO, and some readjustment in the distribution of responsibilities within NATO. As you know, the issue has already surfaced.

Secondly, a unified Europe, which is one of our proclaimed objectives, will insist on a larger voice in keeping with the concept of partnership. Having committed ourselves rhetorically to the idea of a partnership spanning the Atlantic Ocean, are we prepared to give Europe such a larger voice? It is easy to say yes, but that answer has far-reaching implications. Let me name one among many. To give the Europeans an equal voice, as a partner, we would certainly have to give them an equal voice in an area of critical importance to Europe—namely, the Middle East. Are we prepared to share our leadership in the Middle East, and specifically on the Arab-Israeli peace process with the Europeans? The answer in practice is no. In fact, are we prepared to share leadership with Europe more generally? The answer, at best, is ambiguous if one goes beyond the rhetoric. And yet, those are the issues on which we will have to bite the bullet, if we are serious about the fundamental strategic proposition that the larger Europe, but also more unified Europe, is in our national interest. I happen to believe that it is, in the long historical sweep of things, because we cannot indefinitely be simultaneously the leader, and the only truly responsible power in the world. But, if we want others to assume responsibility, we have to share with them some of the decision making. It is a difficult choice.

Making Russia a status quo state is an equally challenging undertaking. It requires the avoidance of antagonism, the restraint on hostility, the furtherance of democracy, and assistance to a country which is economically in a state of disrepute, and dominated by criminalities. It will require a great deal of forbearance, and a broad historical perspective which will enable us to transcend the frustrations and irritations of the moment. We will have to be committed for a long time to come, in helping a Russia which will often appear undeserving of our care, and ungrateful for it. And yet we will have to persist. That persistence will only come with steady, assertive, historically focused leadership. But that is not enough, because you don't obtain someone's collaboration simply by helping him. You also have to create a context in which that collaboration increasingly becomes the only choice that the parties concerned can make.

So, in addition to helping Russia on a long-term basis, and in spite of immediate frustrations, we will very deliberately have to seek a context in which Russia's accommodation with us increasingly becomes their choice. That means creating circumstances in which Russia has no choice but to become a status quo power. That in turn means on the one hand, the expansion of NATO because it does reduce any geopolitical temptations to which Russia at some point may aspire and might be able to exercise even from a position of weakness. On the other hand, it also means creating conditions in the space of the former Soviet Union in which the status quo becomes permanent. That means a deliberate policy of matching aid to Russia with simultaneous aid to the newly independent states of the former Soviet Union. For only if they remain sovereign and independent, will Russia be more inclined to accommodate the status quo society.

Strategically, this means particularly, in my view, focusing on Ukraine. As many of you know that has been my viewpoint for a number of years. I have been propagating this within the Administration, and in this particular instance I think the Administration has adopted the right course of action. It means also choosing several other countries as the foci for our particular attention, irrespective of the degradation of their domestic democratization. It would be nice, of course, if the countries we aid were all brimming with respect for human rights. I would generally prefer that. There may be circumstances, however, in which helping a nondemocratic but newly independent state within the space of the former Soviet Union may, in fact, encourage democracy in Russia.

My choice, in addition to Ukraine, would be Azerbaijan and Uzbekistan, for reasons that are probably familiar to many of you. Uzbekistan because it is the hard core of an independent Central Asia. It is in our interest to preserve an independent Central Asia, because it helps to make Russia a status quo society. Azerbaijan because it is the cork in the bottle. If Azerbaijan is sealed because of Russian, or Russian and Iranian collusion, there is no access for us to Central Asia. Central Asia would become strategically vulnerable. It won't be easy to accomplish this, but I cannot imagine a Western policy which addresses the issue effectively without American leadership.

On the third issue—the Persian Gulf/Middle East—I have already alluded to one prospective issue that we will have to confront: the question of Europe's role. But beyond that there is the question of how do we ensure the stability of the region unless we are prepared to pursue negotiations. The Arab-Israeli peace process is not going to go forward without American leadership. We should have no illusions about that, whatsoever. Whatever progress has been achieved so far, whether it was in the first Sinai disengagement under Nixon and Kissinger, or at Camp David where after thirteen days of intense negotiations, directly led by the President of the United States, in which I personally participated day and night, or in the latter 1980s under Bush and Baker—in each case American leadership was directly and deeply involved. Had it not been for that, there would have been no progress. There would have been no disengagement. There certainly would not have been a Sadat-Begin agreement, and Shamir certainly would have evaded the pressures for peace, if those pressures were confined to those emanating solely from the Arab-Israeli dialogue. It required the United States' insistence. The United States still remains necessary, especially now when the policy of Netanyahu is clearly that of "talk-ing peace, but delaying peace." Pressure on both parties is needed.

Pressure will also be needed on a different issue, one which is very complex and very difficult, but leadership on it is absolutely essential. Namely, in the long run, how sustainable is the policy of dual-containment in the Persian Gulf? What does it accomplish? What are its goals? What is the difference between dual-containment and dual "cop-out?" I find it very difficult to define the difference. Why should we be treating two countries so different from each other as Iraq and Iran under the same rubric, and presumably the same policy? Do we conceivably have some longer term interests with Iran, which it is in our interest to resuscitate, to cultivate, and eventually, to make significant politically? It will require a great deal of sophisticated leadership to move in that direction because the issues are pregnant with domestic political concerns. Yet, in the long run, if we want the region to be stable, I do not see how we can avoid a change in position, and a change in position can only come through leadership.

I don't think I have to belabor the issues pertaining to the Far East. We are all conscious of the fact that really fundamental change is under way. A great new power is in the process of emerging. What it will do, how it will act, and how it will interact with us is clearly going to be a formidable challenge—one which we have not addressed in a consistent fashion. If one compares the course we have pursued over the past three years with respect to Russia with that of our policies toward China, one finds, on a variety of levels, striking contrasts which are difficult to explain. The fact of the matter is that our policy towards China has been contradictory and inadequate. It appears to be devoid of any larger strategic design, and yet such a design is needed. It also is needed because Japan's relationship with us is bound to change. It is, in fact, changing, and it cannot be addressed almost exclusively from the standpoint of trade relations. Thus, here too, a sense of strategic direction requires a great deal of rethinking, then campaigning, articulating, and implementing.

The fifth issue which I mentioned, I deliberately phrased as involving how we manage nuclear proliferation. I did not say how do we stop nuclear proliferation, but how do we manage it. Because it is underway, it has been underway. We have, in fact, in some cases closed our eyes to it, sometimes we have abetted it, and it cannot be stopped.

So the question is, how are we going to live in a world in which nuclear weapons are probably more dispersed, and more available, and where do we draw the effective lines. Is it between different kinds of states, in which case we must more clearly articulate which states are, in our view, entitled to acquire them directly or surreptitiously, and which not? That has been the case so far. We have, in fact, aided some states in attaining nuclear status, even though our policies were proclaimed to be that of nuclear non-proliferation. Or, may we have to draw a line between nation-states, and non-state groupings, particularly terrorist groupings?

It is a fact, though it is an insufficient fact on which to base a policy, that states which have nuclear capabilities have acted with great restraint. Is it possibly the case that states which have an antagonistic relationship with each other become more prudent when both acquire nuclear weapons? Certainly, so far, the Indian-Pakistani confrontation has not been devoid of tension, even the spilling of blood. But it has involved considerable restraint ever since both of them became nuclear-capable. This is an insufficient basis for a grand strategy, but it does suggest, perhaps, that some of our attitudes are hypocritical, and need some rethinking. And again, on this issue American leadership will be of critical significance.

Finally, will large-scale social collapse be avoided? This obviously has a special application for meaning today in Africa. But, this concern can be applied elsewhere as well, in Bosnia which is not exactly the only relevant example. There may be new ones arising, and closer to home. I am far from confident that sociopolitical stability is an enduring reality in Mexico. In any case, large-scale social collapse will pose enormous moral dilemmas for us, and perhaps, in some cases, political challenges.

Zaire is largely a moral dilemma, but should Mexico erupt, or Bosnia again ignite, it would also have a political dimension. Have we provided the leadership that is really in keeping with our posture in the world?

On a crisis of as great a magnitude as the one we are facing in Zaire, it is Canada that is taking the lead, while the Pope and the Secretary-General of the United Nations are appealing for a wider global response, including from the world's only superpower. This will require a degree of commitment and abnegation, and some real sacrifice from us. That is not possible to sustain unless there is a leadership that addresses this issue, speaks to it, and convinces the country that we have a moral, as well as a political interest in addressing this challenge.

In summary, I think the test for us is whether we will prove to be a truly effective, solitary global superpower. Or is there the risk that in shrinking from these challenges, we will be the first impotent global power. And some people are asking the question of whether America is historically fatigued; whether the tricept of power and monopoly and democracy involves an oxymoron.   Perhaps a democracy cannot lead on these issues. Particularly a democracy such as ours, which is becoming increasingly culturally diversified. Under such conditions, a national consensus will be ever more difficult to achieve. I think it is a question certainly worth pondering. Is diversity, as practiced and defined in America today, in fact incompatible with developing and sustaining a national will? For action and leadership has to be derived from national will.

There is also a secondary question. Do we have the structure for decision-making in our society that is responsive to the new global realities? Let me draw your attention to a simple fact, which I know many of you are familiar with. Next year will be the 50th anniversary of the National Security Act. The National Security Act was a belated bureaucratic, institutional reform in response to the inadequacies of our decision-making process during the World War II. It created a great many new innovative processes and procedures, some of which have stood the test of time. Is that machinery adequate today?   Let me cite one specific example which always troubles me. I find it appalling that we don't have any mechanism for effective global political planning in the U.S. government. We do not. There is something called the Policy Planning Council in the Department of State. It has its ups and downs. It has some excellent people on it. But, more often than not, it is a speech-making mechanism for the Secretary of State. That is not altogether bad, because policy is often made by speeches. But, surely, it is not enough.

There are a number of planning mechanisms in the Department of Defense, both in the Secretary's office, and in the Joint Chiefs of Staff. But, you cannot plan national strategy on a complex variety of issues such as the ones I have mentioned from the vantage point of the Defense Department, which involves one particular motivation and perspective. This is not to negate the value of the mechanisms that exist, but they are constrained by a very specific institutional and professional perspective. There is nothing like a global political planning capability in the White House, literally nothing. I find it staggering. I think that the

50th anniversary of the National Security Act suggests that the time has come to remedy this inadequacy.

There is a further problem which concerns me in the background of these. That is with respect to national values and our national culture. It is not simply an unfair charge to assert that our society is becoming an increasingly entertainment-oriented society, that more people than ever before spend more time being mindlessly entertained by procedures and techniques with which you are well familiar. Such a society cannot create and spread competitive ideas that are likely to invoke universal support. At the same time, such societies are likely to produce an increasingly alienated elite that is motivated by contempt for the mass culture, but also driven by disparate power structures.

Today, in a world that is politically inarticulate, effective leadership is impossible without driving ideas behind it. This was the only reason that the Soviet Union was such a powerful state for such a long time. The Soviet Union was always a sham and a front. It hid the reality of poverty, backwardness, and criminality, and yet a great deal came from the power of the ideas, though false, that were identified with the Soviet Union. What are the ideas of our society? These are issues not irrelevant to our future. That is my simple message for this evening.

Part Two

Complexity Theory and National Security Policy

Complex Systems: The Role of Interactions

Robert Jervis

Although we all know that social life and politics constitute systems and that many outcomes are the unintended consequence of complex interactions, the basic ideas of systems do not come readily to mind and so often are ignored. Because I know international politics best and this area is of greatest interest to readers of this book, I will often focus on it. But the arguments are more general and I will take examples from many fields. This is not difficult: systems have been analyzed by almost every academic discipline because they appear throughout our physical, biological, and social world. The fact that congruent patterns can be found across such different domains testifies to the prevalence and power of the dynamics that systems display. Much of this constitutes variations on a few themes, in parallel with

Darwin's summary remark about the structures of living creatures: "Nature is prodigal in variety, but niggard in innovation."[1]

We are dealing with a system when (a) a set of units or elements are inter-connected so that changes in some elements or their relations produce changes in other parts of the system and (b) the entire system exhibits properties and behaviors that are different from those of the parts.

The result is that systems often display non-linear relationships, outcomes cannot be understood by adding together the units or their relations, and many of the results of actions are unintended. Complexities can appear even in what would seem to be simple and deterministic situations. Thus over 100 years ago the mathematician Henri Poincaré showed that the motion of as few as three bodies (such as the sun, the moon, and the earth), although governed by strict scientific laws, defies exact solution: while eclipses of the moon can be predicted thousands of years in advance, they cannot be predicted millions of years ahead, which is a very short period by astronomical standards.[2]

International history is full of inter-connections and complex interactions. Indeed, this one might seem like a parody were it not part of the events leading up to the First World War:

> By the end of the summer of 1913 there
> was a real danger of yet another Balkan
> conflict: the King of Greece [said] that Turkey

was preparing an expedition to recover the island in Greek hands, and from Constantinople the German ambassador reported that the Bulgarian minister to the Porte had informed him of a verbal Turco-Bulgarian agreement under which Bulgaria would attack Thrace in the event of a Turco-Greek war. The danger that a Turco-Greek war could spread beyond the Balkans could not be lightly dismissed. If Turkey and Greece came to blows the Bulgarians could be expected to seek revenge for the defeats of the previous summer; so early a repudiation of the Treaty of Bucharest would offend the Rumanians, whilst the Greeks, if attacked by the Bulgarians, could still invoke their treaty with Serbia. If Serbia became involved no-one could guarantee that Austria-Hungary would once again stand aside.[3]

Ripples move through channels established by actors' interests and strategies. When these are intricate, the ramifications will be as well, and so the results can surprise the actor who initiated the change. The international history of late 19th and early 20th centuries, centered on maladroit German diplomacy, supplies several examples. Dropping the Reinsurance Treaty with Russia in 1890 simplified German diplomacy, as the Kaiser and his advisors had desired. More important, though, were the indirect and delayed consequences, starting with Russia's turn to France, which increased Germany's need for Austrian support, thereby making Germany hostage to her weaker and less stable partner. In 1902, the Germans hoped that the Anglo-Japanese Alliance, motivated by Britain's attempt to reduce her isolation and vulnerability to German pressure, would worsen British relations with Russia (which was Japan's rival in the Far East) and France (which sought British colonial concessions).[4] There were indeed ramifications, but they were not to Germany's liking. The British public became less fearful of foreign ties, easing the way for ententes with France and Russia. Furthermore, Japan, assured of Britain's benevolent neutrality, was able to first challenge and then fight Russia. The Russian defeat, coupled with the strengthening of the Anglo-Japanese treaty, effectively ended the Russian threat to India and so facilitated Anglo-Russian cooperation, much against Germany's interests and expectations.

In a system, the chains of consequences extend over time and many areas: the effects of action are always multiple. Doctors call the undesired impact of medications "side effects." Although the language is misleading—there is no criteria other than our desires that determines which effects are "main" and which are "side"—the point reminds us that disturbing a system will produce several changes. Garrett Hardin gets to the heart of the matter in pointing out that, contrary to many hopes and expectations, we cannot develop or find "a highly specific agent which will do only one thing.... <u>We can never do merely one thing</u>. Wishing to kill insects, we may put an end to the singing of birds. Wishing to 'get there' faster we insult our lungs with smog."[5] Seeking to protect the environment by developing non-polluting sources of electric power, we build windmills that kill hawks and eagles that fly into the blades; cleaning the water in our harbors allows the growth of mollusks and crustaceans that destroy wooden piers and bulkheads; adding redundant safety equipment makes some accidents less likely, but increases the chances of others due to the operators' greater confidence and the interaction effects among the devices; placing a spy in the adversary's camp not only gains valuable information, but leaves the actor vulnerable to deception if the spy is discovered; eliminating rinderpest in East Africa paved the way for canine distemper in lions because it permitted the accumulation of cattle, which required dogs to herd them, dogs which provided a steady source for the virus that could spread to lions; releasing fewer fine particles and chemicals into the atmosphere decreases pollution but also is likely to accelerate global warming; pesticides often destroy the crops that they are designed to save by killing the pests' predators; removing older and dead trees from forests leads to insect epidemics and an altered pattern of regrowth; allowing the sale of an anti-baldness medicine without a prescription may be dangerous because people no longer have to see a doctor, who in some cases would have determined that the loss of hair was a symptom of a more serious problem; flying small formations of planes over Hiroshima to practice dropping the atomic bomb accustomed the population to air raid warnings that turned out to be false alarms, thereby reducing the number of people who took cover on August 6.[6]

In politics, connections are often more idiosyncratic, but their existence guarantees that here too most actions, no matter how well targeted, will have multiple effects. For example, William Bundy was correct to worry that putting troops into Vietnam might not make that country more secure because deployment could not only lead the North to escalate, but also might "(1) cause the Vietnamese government and especially the army to let up [and] (2) create adverse public reactions to our whole presence on 'white men' and 'like the French' grounds."[7] It seems that the American development of nuclear weapons simultaneously restrained Stalin by increasing his fear of war and made him "less cooperative and less willing to compromise, for fear of seeming weak."[8] Indeed, it is now widely accepted that mutual second strike capability not only decreased the chance of nuclear war but also made it safer for either side to engage in provocations at lower levels of violence.[9] (Similarly, providing security guarantees to the countries of East Europe might lead them to take harsher stances toward minority ethnic groups and make fewer efforts to maintain good relations with their neighbors.) To mention three more surprising cases, in the fall of 1948 General Clay warned that American budget deficits would be seen in Europe as a forerunner of inflation and so would undermine morale in West Berlin; the American pressure on the Europeans to rearm more rapidly in response to the North Korean attack on the South produced squabbles that encouraged the USSR "to believe that contradictions in the enemy camp ultimately would tear apart the enemy coalition....[and so] undermined U.S. bargaining power"; in 1994 the dollar strengthened after President Clinton hired a powerful lawyer to defend him against charges of sexual harassment: as one currency trader put it, "we were starting to lose faith in him and that helped turn things."[10]

Interactions, Not Additivity

Because of the prevalence of inter-connections, we cannot understand systems by summing the charac-teristics of the parts or the bilateral relations between pairs of them.[11] This is not to say that such operations are never legitimate, but only that when they are we are not dealing with a system. More precisely, ac-tions often interact to produce results that cannot be comprehended by linear models.

Linearity involves two propositions: (1) changes in sys-tem output are proportional to changes in input...and (2) system outputs corresponding to the sum of two inputs are equal to the sum of the outputs arising from the individual inputs.[12]

Intuitively, we often expect linear relationships. If a little foreign aid slightly increases economic growth, then more aid should produce greater growth. But in a system a variable may operate through a non-linear function. That is, it may have a disproportionate im-pact at one end of its range. Sometimes even a small amount of the variable can do a great deal of work and then the law of diminishing returns sets in, as is often the case for the role of catalysts. In other cases very little impact is felt until a critical mass is assembled. For example, women may thrive in a profession only after there are enough of them so that they do not feel like strangers. Clausewitz noted a related effect:

> The scale of a victory does not increase simply at a rate commensurate with the increase in size of the defeated armies, but progressively. The outcome of a major battle has a greater psychological effect on the loser than the winner. This, in turn, gives rise to additional loss of material strength [through abandonment of weapons in a retreat or desertions from the army], which is echoed in loss of morale; the other two become mutually interactive as each enhances and intensifies the other.[13]

Similarly, the effect of one variable or characteristic can depend on which others are present. Thus even if it is true that democracies do not fight each other in a world where other regimes exist, it would not follow that an entirely democratic world would necessarily be a peaceful one: democracies might now be united by opposition to or the desire to be different from autocracies and once triumphant might turn on each other. (The other side of this coin is that many of the characteristics of democracies that classical Realists saw as undermining their ability to conduct foreign policy—the tendency to compromise, heed public opinion, and assume others are reasonable—may serve them well when most of their interactions are with other democracies.)

To further explore interactions it is useful to start with the basic point that the results cannot be predicted from examining the individual inputs separately. I will then move on to the ways in which the effect of one actor's strategy depends on that of others, after which I will discuss how the actors and their environments shape each other, sometimes to the point where we should make the interaction itself the unit of analysis.

First Interactions: Results Cannot Be Predicted From the Separate Actions

The effect of one variable frequently depends on the state of another, as we often see in everyday life: each of two chemicals alone may be harmless but exposure to both could be fatal; patients have suffered from taking combinations of medicines that individually are helpful. So research tries to test for interaction effects and much of modern social science is built on the understanding that social and political outcomes are not simple aggregations of the actors' preferences because very different results are possible depending on how choices are structured and how actors move strategically.

Turning to international politics, Shibley Telhami argues that while pan-Arabism and pro-Palestinian sentiment worked to enhance Egyptian influence when Egypt was strong, they made it more dependent on other Arab states when Egypt was weak.[14] From the fact—if it is a fact—that nuclear weapons stabilized Soviet-American relations we cannot infer that they would have a similar impact on other rivalries because variables that interact with nuclear weapons may be different in these cases (and of course may vary from one pair of rivals to another). Within the military domain one finds interaction effects as well: two weapons or tactics can work particularly well together and indeed most analysts stress the value of "combined arms" techniques that coordinate the use of infantry, artillery, armor, and aircraft. Events that occur close together also can have a different impact than they would if their separate influences were merely summed. The Soviet invasion of Afghanistan affected American foreign policy very deeply in part because it came on the heels of the Iranian revolution, which undercut American power, disturbed public opinion, and frightened allies.

In explaining outcomes, we are prone to examine one side's behavior and overlook the stance of the other with which it is interacting. Although deterrence theory is built on the idea of interdependent decisions, most explanations for why deterrence succeeds in some cases and fails in others focus on differences in what the defender did while ignoring variation in the power and motivation of the challenger, just as much policy analysis in general starts—and often ends—with the strengths and weaknesses of the policies contemplated and adopted. But one hand cannot clap; we need to look at the goals, resources, and policies of those with whom the actor is dealing. Teachers are prone to make the parallel error of not exploring how shortcomings in our students' performances on tests may be attributable to the questions we ask.

Second Interactions: Strategies Depend on the Strategies of Others

Further complexities are introduced when we look at the interactions that occur between strategies when actors consciously react to others and anticipate what they think others will do. Obvious examples are provided by many diplomatic and military surprises: a state believes that the obstacles to a course of action are so great that the adversary could not undertake it; the state therefore does little to block or prepare for that action; the adversary therefore works especially hard to see if he can make it succeed. As an 18th century general explained, "In war it is precisely the things which are thought impossible which most often succeed, when they are well conducted."[15] In the war in Vietnam, the U.S. Air Force missed this dynamic and stopped patrolling sections of the North's supply lines when reconnaissance revealed that the number of targets had greatly diminished: after the attacks ceased the enemy resumed use of the route.[16]

Both the success and failures of policies are determined interactively. This means that many cases of intelligence failure are mutual—i.e., they are failures by the side that took the initiative as well as by the state that was taken by surprise. Indeed, an actor's anticipation of what others will do stems in part from its estimate of what the other thinks the actor will do. In many cases of surprise a state sees that a certain move by the adversary cannot succeed and therefore does not expect the other to take it: the U.S. did not expect the Russians to put missiles into Cuba or Japan to attack Pearl Harbor because American officials knew that the U.S. would thwart these measures if they were taken. These judgments were correct, but because the other countries saw the world and the U.S. less accurately, the American predictions were also inaccurate.[17]

Third Interactions: Behavior Changes the Environment

Initial behaviors and outcomes often influence later ones, producing powerful dynamics that explain change over time and that cannot be captured by labeling one set of elements "causes" and other "effects." Although learning and thinking play a large role in political and social life, they are not necessary for this kind of temporal interaction. Indeed, it characterizes the operation of evolution in nature. We usually think of individuals and species competing with one another within the environment, thus driving evolution through natural selection. In fact, however, there is coevolution: plants and animals not only adapt to the environment, they change it. As a result, it becomes more hospitable to some life forms and less hospitable to others.

Nature is not likely to "settle down" to a steady state as the development or growth of any life form will consume—and be consumed by—others, closing some ecological niches and opening others, which in turn will set off further changes. To some extent, organisms create their own environments, not only by direct actions (e.g., digging burrows, storing food, excreting waste products), but as their very existence alters the microclimates, nutrients, and feeding opportunities that will affect them and others. Indeed, not only does the amount of rainfall influence the vegetation that grows, but the latter affects the former as well. To take a more readily visible example, elephants thrive on acacia trees. But the latter can only develop in the absence of the former. After a while, the elephants destroy the trees, drastically changing the wildlife that the area can sustain and even affecting the physical shape of the land. In the process, they render the area uncongenial to themselves, and they either die or move on. The land is adapting to the elephants just as they are to it. One Maasai put it well: "Cows grow trees, elephants grow grasslands."[18] Most consequentially, the very atmosphere that supports current life was produced by earlier forms, many of which could not survive in the new environment: long before humans, species of bacteria were so successful and generated so much pollution that they poisoned themselves.

Politics, like nature, rarely settles down as each dispute, policy, or action affects others and re-shapes the political landscape, inhibiting some behaviors and enabling others. Campaign financing reforms generated new actors in the form of PACs, new issues in the form of arguments about what PAC activities should be permitted, new debates about the meaning of the first amendment, and new groups that track the flow of money and services. These in turn affect not only how the funds are solicited and given, but also change the allies and adversaries that are available to political actors and the ways in which a variety of other issues are thought of. Political maneuvers create niches for new actors and disputes, often in ways that no one had anticipated. William Miller's fascinating study of the Southern attempt to control—indeed choke off—the debate about slavery in the 1830s points out that by passing a "gag rule" prohibiting Congressional discussion of petitions asking for the end of the slave trade in the District of Columbia, the South called up "petitions against the gag rule itself" and made a new issue of the right to petition the government.[19] Indeed, many protest movements grow as people previously unsympathetic are offended by the way the authorities respond. Each added issue may mobilize the population in a different way than did the original one—and of course the new dispute in turn changes the political environment.

It is clear that, for better and for worse, people change as they are affected by experiences, including those that they have chosen. Personal development does not mean that the person simply turns into what was latent in him. Instead, we need to take account of the situations in which he was placed. To take some examples familiar to academics, when we think about whether one of our bright undergraduates would do well in a Ph.D. program, we are likely to ask whether she enjoys and does well at independent research. But the right question may be whether she will enjoy and do well at it after she has experienced two or three years of graduate school. It is also a mistake to point to the lackluster career of a person who failed to get into a major graduate school or to receive tenure at a top school as justification for these decisions because we do not know how well she would have done in a more stimulating and demanding setting. Similar reasoning explains the limitations of the common argument that international institutions do not matter because states will ignore them "when push comes to shove" and vital interests are at stake. Although the statement is correct, it misses the role institutions can play in shaping interests and seeing that push does not come to shove.

Many of the ways in which deterrence can fail reflect interactions in which the state's behavior changes its environment. Thus it might seem obvious that for Pakistan to build nuclear weapons could not but decrease the likelihood of an Indian attack. But this would overlook both the danger that India would feel increased pressures to preempt and the likely Indian judgment that world public opinion would be less censorious of an attack against a nuclear-armed Pakistan. Furthermore, the deterrence tactics that bring success at one point can change the other side and make future deterrence more difficult, most obviously by increasing the challenger's dissatisfaction with the status quo and giving it incentives to "design around" the defender's threats that previously had been adequate.[20] A version of this process may have been at work in Vietnam in the late 1950s, casting a somewhat different light on the standard account of the American advisors erroneously fearing a conventional attack from the North and so training the South Vietnamese army to meet a fictitious danger. But the reason Diem's enemies turned to guerrilla warfare may have been that he had succeeded in foreclosing the option of fighting a conventional war. The American policy was still in error in failing to anticipate the response it could trigger, but not in having been misconceived from the start.

Because actions change the environment in which they operate, identical but later behavior does not produce identical results: history is about the changes produced by previous thought and action as people and organizations confront each other through time. The final crisis leading to World War II provides an illustration of some of these processes. Hitler had witnessed his adversaries give in to pressure; as he explained, "Our enemies are little worms. I saw them at Munich."[21] But the allies had changed because of Hitler's behavior. So had Poland. As A.J.P. Taylor puts it, "Munich cast a long shadow. Hitler waited for it to happen again; Beck took warning from the fate of Benes."[22]

Hitler was not the only leader to fail to understand that his behavior would change his environment. Like good linear social scientists, many statesmen see that their actions can produce a desired outcome, all other things being equal, and project into the future the maintenance of the conditions that their behavior will in fact undermine. This in part explains the Argentine calculations preceding the seizure of the Falklands/ Malvinas. Their leaders could see that Britain's ability to protect its position was waning, as evinced by the declining naval presence, and that Argentina's claim to the islands had received widespread international support. But what they neglected was the likelihood that the invasion would alter these facts, unifying British opinion against accepting humiliation and changing the issue for international audiences from the illegitimacy of colonialism to the illegitimacy of the use of force. A similar neglect of the transformative power of action may explain why Saddam Hussein thought he could conquer Kuwait. Even if America wanted to intervene, it could do so only with the support and cooperation of other Arab countries, which had sympathized with Iraq's claims and urged American restraint. But the invasion of Kuwait drastically increased the Arabs' perception of threat and so altered their stance. Furthermore, their willingness to give credence to Iraqi promises was destroyed by the deception that had enabled the invasion to take everyone by surprise. Germany's miscalculation in 1917 was based on a related error: although unrestricted submarine warfare succeeded in sinking more British shipping than the Germans had estimated would be required to drive Britain from the war, the American entry (which Germany expected) led the British to tolerate shortages that otherwise would have broken their will because they knew that if they held out, the U.S. would rescue them.[23]

The failure to appreciate the fact that the behavior of the actors is in part responsible for the environment which then impinges on them can lead observers— and actors as well—to underestimate actors' influence. Thus states caught in a conflict spiral believe that they have little choice but to respond in kind to the adversary's hostility. This may be true, but it may have been the states' earlier behavior that generated the situation that now is compelling. Robert McNamara complains about how he was mislead by faulty military reporting but similarly fails to consider whether his style and pressure might have contributed to what he was being told.[24]

Products of Interaction as the Unit of Analysis

Interaction can be so intense and transformative that we can no longer fruitfully distinguish between actors and their environments, let alone say much about any element in isolation. We are accustomed to referring to roads as safe or dangerous, but if the drivers understand the road conditions this formulation may be misleading: the knowledge that, driving habits held constant, one stretch is safe or dangerous will affect how people drive—they are likely to slow down and be more careful when they think the road is dangerous and speed up and let their attention wander when it is "safe." It is then the road-driver system that is the most meaningful unit of analysis. In the wake of the sinking of a roll-on roll-off ferry, an industry representative said:

> With roro's, the basic problem is that you have a huge open car deck with doors at each end. But people are well aware of this, and it is taken into account in design and operation. You don't mess around with them. There have not been too many accidents because they are operated with such care.[25]

Similarly, we often refer to international situations as precarious, unstable, or dangerous. But, again, if statesmen perceive them as such and fear the consequences, they will act to reduce the danger— one reason why the Cuban missile crisis did not lead to war was that both sides felt that this could be the outcome if they were not very careful. Nuclear weapons generally have this effect. Because statesmen dread all-out war, international politics is safer than it would otherwise be, and probably safer than if war were less destructive. Conversely, like drivers on a "safe" stretch of road, decision-makers can behave more recklessly in calmer times because they have more freedom to seek unilateral gains as well as needing to generate risk to put pressure on others. For example, the relaxation of Anglo-German tensions after 1911 may have misled both countries into believing that they could afford dangerous tactics in 1914.

Circular Effects

Systems can produce circular effects as actors respond to the new environments their actions have created, often changing themselves in the process. In international politics, perhaps the most important manifestation of this dynamic is the large-scale operation of the security dilemma—i.e., the tendency for efforts to increase a state's security to simultaneously decrease the security of others. Because states know that they cannot rely on others in the unpredictable future, they seek to protect themselves against a wide range of menaces. Thus in the 1930s Japan, which was heavily dependent on resources from outside its borders, sought to expand the area it controlled. Immediate economic needs generated by the world-wide depression increased but did not create this impulse. Nor were they brought on by specific conflicts with the Western powers. Rather what was driving was the fear that conflict might be forced upon Japan in the future, which meant that to remain secure Japan needed raw materials and larger markets. The result was the conquest of Manchuria, followed by a larger war with China, and then by the occupation of Indochina. Each move generated resistance that made the next action seem necessary, and the last move triggered the American oil embargo, which in turn pushed Japan into attacking the West before it ran out of oil. Had Japan been secure, her aggression would not have been necessary; it was the fear of an eventual war with the West that required policies that moved Western enmity from a possibility to a reality. (Of course a further irony is that World War II led to the reconstruction of international politics and the Japanese domestic system that brought Japan security, economic dominance of South East Asia, and access to markets around the world.)

Despite the familiarity of the idea that social action forms and takes place within a system, scholars and statesmen as well as the general public are prone to think in non-systemic terms. This is often appropriate, and few miracles will follow from thinking systemically because the interactive, strategic, and contingent nature of systems limits the extent to which complete and deterministic theories are possible. But we need to take more seriously the notion that we are in a system and to look for the dynamics that drive them. A distinguished student of genetics summarized his perspective in the phrase: "Nothing in biology makes sense except in the light of evolution."[26] Very little in social and political life makes sense except in the light of systemic processes. Exploring them gives us new possibilities for understanding and effective action; in their absence we are likely to flounder.

End Notes

1. Charles Darwin, *The Origin of Species* (New York: Modern Library, 1936), p. 143.

2. For a recent discussion, see Robert Pool, "Chaos Theory: How Big an Advance?" *Science*, vol. 245, July 9, 1989, p. 26.

3. R.J. Crampton, *The Hollow Detente: Anglo-German Relations in the Balkans, 1911-1914* (Atlantic Highlands, N.J.: Humanities Press, 1980), p. 131.

The first scholars who applied ideas from ecology to international politics noted that when elements are interconnected, "any substantial change in one sector of the milieu is nearly certain to produce significant, often unsettling, sometimes utterly disruptive consequences in other sectors." (Harold and Margaret Sprout, *An Ecological Paradigm for the Study of International Politics* (Princeton University, Center for International Studies, Research Memorandum no. 30, March 1968), p. 55.) A more recent study in this vein is James Rosenau, *Turbulence in World Politics* (Princeton: Princeton University Press, 1990).

4. P.J.V. Rolo, *Entente Cordiale* (New York: St. Martin's Press, 1969), p. 121.

5. Garrett Hardin, "The Cybernetics of Competition," *Perspectives in Biology and Medicine*, vol. 7, Autumn 1963, pp. 79-80, emphasis added.

6. Jonathan Weisman, "Tilting At Windmills," *Wildlife Conservation*, vol. 97, January/February 1994, pp. 52-57; Lindsey Gruson, "Problem With Clean Harbor: Creatures Devour Waterfront," *New York Times*, June 27, 1993; Aaron Wildavsky, *Searching for Safety* (New Brunswick, N.J.: Transaction Books, 1988); Perrow, *Normal Accidents*; the classic case of "turned" agents was revealed in J.C. Masterman, *The Double-Cross System in the War of 1939 to 1945* (New Haven: Yale University Press, 1972); Packer, "Coping with a Lion Killer," pp. 14-17; William Stevens, "Acid Rain Efforts Found to Undercut Themselves," *New York Times*, January 27, 1994; Richard Kerr, "Study Unveils Climate Cooling Caused by Pollutant Haze," *Science*, vol. 268, May 12, 1995, p. 802; Kerr, "It's Official: First Glimmer of Greenhouse Warning Seen," ibid, vol. 270, December 8, 1995, pp. 1565-67; Nancy Langston, *Forest Dreams, Forest Nightmares: The Paradox of Old Growth in the InlandWest* (Seattle: University of Washington Press, 1995), pp. 148-50; 292-94; "You Want Hair, Get A Prescription," *Aspen Daily News*, July 28, 1994 (in the end, the FDA decided to permit freer sale of the medication: "Hair-Growth Drug to Be Sold Over the Counter," *New York Times*, February 13, 1996, p. C 10; Leon Sigal, *Fighting to a Finish: The Politics of War Termination in the United States and Japan, 1945* (Ithaca: Cornell University Press, 1988), pp. 215-16. Those who believe that it is healthy to eat food that has not been treated with pesticides will be interested in Jane Brody, "Strong Views on Origins of Cancer,"

New York Times, July 5, 1994. But as we will discuss in chapters 2 and 7, not all unintended consequences are undesired.

7. Quoted in Larry Berman, "Coming to Grips with Lyndon Johnson's War," *Diplomatic History*, vol. 17, Fall 1993, p. 525.

8. David Holloway, *Stalin and the Bomb* (New Haven: Yale University Press, 1994), p. 272.

9. Glenn Snyder, "The Balance of Power and the Balance of Terror," in Paul Seabury, ed., *The Balance of Power* (San Francisco: Chandler, 1965), pp. 184-201.

10. Walter Millis, ed., *The Forrestal Diaries* (New York: Viking, 1951), p. 526; William Stueck, *The Korean War* (Princeton: Princeton University Press, 1995), p. 6; quoted in Thomas Friedman, "It's a Mad, Mad, Mad, Mad, World Money Market," *New York Times*, May 8, 1994, section E, p. 2. As these examples show, people's expectations—based in part on their beliefs about others' expectations—are central to the dynamics of systems.

11. Kenneth Waltz, *Theory of International Politics* (Reading, Mass.: Addison-Wesley, 1979), p. 64; for parallel discussions in social psychology, organization theory, and ecology, see respectively Paul Watzlawick, Janet Beavin, and Don Jackson, *Pragmatics of Human Communication: A Study of Interactional Patterns, Pathologies, and Paradoxes* (New York: Norton, 1967), pp. 125-26, 135-39; Charles Perrow, *Normal Accidents* (New York: Basic Books, 1984); Pimm, *Balance of Nature?*, pp. 249-56.

12. Alan Beyerchen, "Nonlinear Science and the Unfolding of a New Intellectual Vision," in Richard Bjornson and Marilyn Waldman, eds., *Papers in Comparative Studies* vol. 6 (Columbus, Ohio: Center for Comparative Studies in the Humanities, Ohio State University Press, 1989), p. 30.

13. Carl von Clausewitz, *On War*, ed. and trans. Michael Howard and Peter Paret (Princeton: Princeton University Press, 1976), p. 253. For a general discussion of war from the perspective of systems effects, see Roger Beaumont, *War, Chaos, and History* (Westport, Conn.: Praeger, 1994).

14. Shibley Telhami, *Power and Leadership in International Bargaining: The Path to the Camp David Accords* (New York: Columbia University Press, 1990), pp. 12, 92-106.

15. Quoted in Reed Browning, *The War of the Austrian Succession* (New York: St. Martin's Press, 1993), p. 123.

16. Barry Watts, "Unreported History and Unit Effectiveness," *Journal of Strategic Studies*, vol. 12, March 1989, p. 98.

17. Klaus Knorr, "Failures in National Intelligence Estimates: The Case of the Cuban Missiles," *World Politics*, vol. 16, April 1964, pp. 455-67; for a related discussion, see James Wirtz, *The Tet Offensive* (Ithaca: Cornell University Press, 1991).

18. Quoted in David Western, "The Balance of Nature," *Wildlife Conservation*, vol. 96, March/April 1993, p. 54.

19. William Lee Miller, *Arguing About Slavery* (New York: Knopf, 1996), p. 278. The classic treatments of these processes are E.E. Schattschneider, *Politics, Pressures and*

the Tariff (New York: Prentice-Hall, 1935) and Schattschneider, *The Semisovereign People* (New York: Holt, Rinehart, and Winston, 1960).

20. Alexander George and Richard Smoke, *Deterrence in American Foreign Policy* (New York: Columbia University Press, 1974), pp. 522-30.

21. Quoted in Glenn Snyder and Paul Diesing, *Conflict Among Nations* (Princeton: Princeton University Press, 1977), p. 187.

22. A.J.P. Taylor, *The Origins of the Second World War* (Greenwich, Conn.: Fawcett, 1966), p. 242.

23. Fred Charles Iklé, *Every War Must End* (New York: Columbia University Press, 1971), pp. 42-48.

24. Robert McNamara, *In Retrospect: The Tragedy and Lessons of Vietnam* (New York: Times Books, 1995); Deborah Shapley, *Promise and Power: The Life and Times of Robert McNamara* (Boston: Little Brown, 1993), pp. 149-52. "`Ah, les statistiques!'* one of the Vietnamese generals exclaimed to an American friend. `Your Secretary of Defense loves statistics. We Vietnamese can give him all he wants. If you want them to go up, they will go up. If you want them to go down, they will go down'": Roger Hilsman, *To Move A Nation* (Garden City, N.Y.: Doubleday, 1967), p. 523.

25. Quoted in Stephen Kenzer, "Little Hope for 800 Lost in Sinking of Baltic Sea Ferry," *New York Times*, September 29, 1994.

26. Theodosius Dobzhansky, "Nothing in Biology Makes Sense Except in the Light of Evolution," *American Biology Teacher*, vol. 35, March 1972, pp. 125-29.

Many Damn Things Simultaneously: Complexity Theory and World Affairs[1]

James N. Rosenau

In this emergent epoch of multiple contradictions that I have labeled "fragmegration" in order to summarily capture the tensions between the fragmenting and integrating forces that sustain world affairs,[2] a little noticed—and yet potentially significant—discrepancy prevails between our intellectual progress toward grasping the underlying complexity of human systems and our emotional expectation that advances in complexity theory may somehow point the way to policies which can ameliorate the uncertainties inherent in a fragmegrative world. The links here are profoundly causal: the more uncertainty has spread since the end of the Cold War, the more are analysts inclined to seek panaceas for instability and thus the more have they latched onto recent strides in complexity theory in the hope that it will yield solutions to the intractable problems that beset us. No less important, all these links—the uncertainty, the search for panaceas, and the strides in complexity theory—are huge, interactive, and still intensifying, thus rendering the causal dynamics ever more relevant to the course of events.

In short, all the circumstances are in place for an eventual disillusionment with complexity theory. For despite the strides, there are severe limits to the extent to which such theory can generate concrete policies that lessen the uncertainties of a fragmegrated world. And as these limits become increasingly evident subsequent to the present period of euphoria over the theory's potential utility, a reaction against it may well set in and encourage a reversion back to simplistic, either/or modes of thought. Such a development would be regrettable. Complexity theory does have insights to offer. It provides a cast of mind that can clarify, that can alert observers to otherwise unrecognized problems, and that can serve as a brake on undue enthusiasm for particular courses of action. But these benefits can be exaggerated and thus disillusioning. Hence the central purpose of this paper is to offer a layman's appraisal of both the potentials and the limits of complexity theory—to differentiate what range

of issues and processes in world affairs it can be reasonably expected to clarify from those that are likely
to remain obscure.

Uncertainties

That a deep sense of uncertainty should pervade
world affairs since the end of the Cold War is hardly
surprising. The U.S.-Soviet rivalry, for all its tensions
and susceptibility to collapsing into nuclear holocaust,
intruded a stability into the course of events that was
comprehensible, reliable, and continuous. The enemy
was known. The challenges were clear. The dangers
seemed obvious. The appropriate responses could
readily be calculated. Quite the opposite is the case
today, however. If there are enemies to be contested,
challenges to meet, dangers to avoid, and responses
to be launched, we are far from sure what they are.
So uncertainty is the norm and apprehension the
mood. The sweet moments when the wall came down
in Berlin, apartheid ended in South Africa, and an aggression was set back in Kuwait seem like fleeting
and remote fantasies as the alleged post-Cold War
order has emerged as anything but orderly. Whatever may be the arrangements that have replaced the
bipolarity of U.S.-Soviet rivalry, they are at best incipi-
ent structures and, at worst, they may simply be
widespread disarray.

Put differently, a new epoch can be said to be evolving. As indicated, it is an epoch of multiple
contradictions: The international system is less dominant, but it is still powerful. States are changing, but

they are not disappearing. State sovereignty has eroded, but it is still vigorously asserted. Governments are weaker, but they can still throw their weight around. At times publics are more demanding, but at other times they are more compliant. Borders still keep out intruders, but they are also more porous. Landscapes are giving way to ethnoscapes, mediascapes, ideoscapes, technoscapes, and finanscapes, but territoriality is still a central preoccupation for many people.[3]

Sorting out contradictions such as these poses a number of difficult questions: How do we assess a world pervaded with ambiguities? How do we begin to grasp a political space that is continuously shifting, widening and narrowing, simultaneously undergoing erosion with respect to many issues and reinforcement with respect to other issues? How do we reconceptualize politics so that it connotes identities and affiliations as well as territorialities? How do we trace the new or transformed authorities that occupy the new political spaces created by shifting and porous boundaries?

The cogency of such questions—and the uncertainty they generate—reinforce the conviction that we are deeply immersed in an epochal transformation sustained by a new world view about the essential nature of human affairs, a new way of thinking about how global politics unfold. At the center of the emergent world view lies an understanding that the order which sustains families, communities, countries, and the world through time rests on contradictions, ambiguities, and uncertainties. Where earlier epochs were conceived in terms of central tendencies and orderly patterns, the present epoch appears to derive its order from contrary trends and episodic patterns. Where the lives of individuals and societies were once seen as moving along linear and steady trajectories, now their movement seems nonlinear and erratic, with equilibrium being momentary and continuously punctuated by sudden accelerations or directional shifts.

Accordingly, the long-standing inclination to think in either/or terms has begun to give way to framing challenges as both/and problems. People now understand, emotionally as well as intellectually, that unexpected events are commonplace, that anomalies are normal occurrences, that minor incidents can mushroom into major outcomes, that fundamental processes trigger opposing forces even as they expand their scope, that what was once transitional may now be enduring, and that the complexities of modern life are so deeply rooted as to infuse ordinariness into the surprising development and the anxieties that attach to it.

To understand that the emergent order is rooted in contradictions and ambiguities, of course, is not to lessen the sense of uncertainty as to where world affairs are headed and how the course of events is likely to impinge on personal affairs. Indeed, the more one appreciates the contradictions and accepts the ambiguities, the greater will be the uncertainty one experiences. And the uncertainty is bound to intensify the more one ponders the multiplicity of reasons why the end of the Cold War has been accompanied by pervasive instabilities. Clearly, the absence of a superpower rivalry is not the only source of complexity. Technological dynamics are also major stimulants, and so are the breakdown of trust, the shrinking of distances, the globalization of economies, the explosive proliferation of organizations, the information revolution, the fragmentation of groups, the integration of regions, the surge of democratic practices, the spread of fundamentalism, the cessation of intense enmities, and the revival of historic animosities—all of which in turn provoke further reactions that add to the complexity and heighten the sense that the uncertainty embedded in nonlinearity has become an enduring way of life.

In some corners of the policy-making community there would appear to be a shared recognition that the intellectual tools presently available to probe the pervasive uncertainty underlying our emergent epoch may not be sufficient to the task. More than a few analysts could be cited who appreciate that our conceptual equipment needs to be enhanced and refined, that under some conditions nonlinear approaches are more suitable than the linear conceptual equipment that has served for so long as the basis of analysis, that the disciplinary boundaries that have separated the social sciences from each other and from the hard sciences are no longer clear-cut, and that the route to understanding and sound policy initiatives has to be traversed through interdisciplinary undertakings.[4]

It is perhaps a measure of this gap between the transformative dynamics and the conceptual equipment available to comprehend them that our vocabulary for understanding the emergent world lags well behind the changes themselves. However messy the world may have been in the waning epoch, at least we felt we had incisive tools to analyze it. But today we still do not have ways of talking about the diminished role of states without at the same time privileging them as superior to all the other actors in the global arena. We lack a means for treating the various contradictions as part and parcel of a more coherent order. We do not have techniques for analyzing the simultaneity of events such that the full array of their interconnections and feedback loops are identified.

Searching for Panaceas

So it is understandable that both the academic and policy-making communities are vulnerable to searching for panaceas. Aware they are ensconced in an epoch of contradictions, ambiguities, and uncertainties, and thus sensitive to the insufficiency of their conceptual equipment, officials and thoughtful observers alike may be inclined to seek security through an overall scheme that seems capable of clarifying the challenges posed by the emergent epoch. Complexity theory is compelling in this regard. The very fact that it focuses on complex phenomena and presumes that these are subject to theoretical inquiry, thereby implying that complex systems are patterned and ultimately comprehensible, may encourage undue hope that humankind's problems can be unraveled and effective policies designed to resolve them pursued.

Stirring accounts of The Santa Fe Institute, where complexity theory was nursed into being through the work
of economists, statisticians, computer scientists, mathematicians, biologists, physicists, and political
scientists in a prolonged and profoundly successful
interdisciplinary collaboration, kindled these hopes.[5]
The stories of how Brian Arthur evolved the notion of
increasing returns in economics, of how John H. Holland developed genetic algorithms that could result in
a mathematical theory capable of illuminating a wide
range of complex adaptive systems, of how Stuart
Kauffman generated computer simulations of abstract,
interacting agents that might reveal the inner workings of large, complicated systems such as the United
States, of how Per Bak discovered self-organized criticality that allowed for inferences as to how social
systems might enter upon critical states that jeopar-
dize their stability, of how Murray Gell-Mann pressed
his colleagues to frame the concept of co-evolution
wherein agents interact to fashion complex webs of
interdependence—these stories suggested that
progress toward the comprehension of complex systems was bound to pay off. And to add to the sense of
panaceas, expectations were heightened by the titles
these scholars gave to their works written to make
their investigations meaningful for laymen. Consider,
for example, the implications embedded in Holland's
Hidden Order[6] and Kauffman's *At Home in the Universe*[7] that creative persistence is worth the effort in
the sense that eventually underlying patterns, a hidden order, are out there to be discovered.[8]

There are, in short, good reasons to be hopeful: if those on the cutting edge of inquiry can be sure that human affairs rest on knowable foundations, surely there are bases for encouragement that the dilemmas of the real, post-cold war world are susceptible to clarification and more effective control. Never mind that societies are increasingly less cohesive and boundaries increasingly more porous; never mind that vast numbers of new actors are becoming relevant to the course of events; never mind that money moves instantaneously along the information highway and that ideas swirl instantaneously in cyberspace; and never mind that the feedback loops generated by societal breakdowns, proliferating actors, and boundary-spanning information are greatly intensifying the complexity of life late in the 20th Century—all such transformative dynamics may complicate the task of analysts, but complexity theory tells us that they are not beyond comprehension, that they can be grasped.

I do not say this sarcastically. Rather, I accept the claims made for complexity theory. It has made enormous strides and it does have the potential for clarifying and ultimately ameliorating the human condition. Its progress points to bases for analytically coping with porous boundaries, societal breakdowns, proliferating actors, fast-moving money and ideas, and elaborate feedback loops. But to stress these strides is not to delineate a time line when they will reach fruition in terms of policy payoffs, and it is here, in the discrepancy between the theoretical strides and their policy relevance, that the need to highlight theoretical limits and curb panacean impulses arises.

Strides in Complexity Theory

Before specifying the limits of complexity theory, let us first acknowledge the claims made for it. This can be accomplished without resort to mathematical models or sophisticated computer simulations. Few of us can comprehend the claims in these terms, but if the theoretical strides that have been made are assessed from the perspective of the philosophical underpinnings of complexity theory, it is possible to identify how the theory can serve the needs of those of us in the academic and policy-making worlds who are not tooled up in mathematics or computer science but who have a felt need for new conceptual equipment. Four underpinnings of the theory are sufficient for this purpose. The four are equally important and closely interrelated, but they are briefly outlined separately here in order to facilitate an assessment of the theory's relevance to the analysis of world affairs.

As I understand it, at the core of complexity theory is the complex adaptive system—not a cluster of unrelated activities, but a system; not a simple system, but a complex one; and not a static, unchanging set of arrangements, but a complex adaptive system. Such a system is distinguished by a set of interrelated parts, each one of which is potentially capable of being an autonomous agent that, through acting autonomously, can impact on the others, and all of which either engage in patterned behavior as they sustain day-to-day routines or break with the routines when new challenges require new responses and new patterns. The interrelationships of the agents is what makes them a system. The capacity of the agents to break with routines and thus initiate unfamiliar feedback processes is what makes the system complex (since in a simple system all the agents consistently act in prescribed ways.) The capacity of the agents to cope collectively with the new challenges is what makes them adaptive systems. Such, then, is the modern urban community, the nation state, and the international system. Like any complex adaptive system in the natural world, the agents that comprise world affairs are brought together into systemic wholes that consist of patterned structures ever subject to transformation as a result of feedback processes from their external environments or from internal stimuli that provoke the agents to break with their established routines. There may have been long periods of stasis in history where, relatively speaking, each period in the life of a human system was like the one before it, but for a variety of reasons elaborated elsewhere,[9] the present period is one of turbulence, of social systems and their polities undergoing profound transformations that exhibit all the characteristics of complex adaptive systems.

The four premises of complexity theory build upon this conception. They call attention to dimensions of complex adaptive systems that both offer promising insights into world affairs and highlight the difficulties of applying complexity theory to policy problems.

Self-Organization and Emergent Properties

The parts or agents of a complex adaptive system, being related to each other sufficiently to form recurrent patterns, do in fact self-organize their patterned behavior into an orderly whole[10] and, as they do, they begin to acquire new attributes. The essential structures of the system remain intact even as their emergent properties continue to accumulate and mature. Through time the new properties of the system may obscure its original contours, but to treat these processes of emergence as forming a new system is to fail to appreciate a prime dynamic of complexity, namely, the continuities embedded in emergence. As one analyst puts it, the life of any system, "at all levels, is not one damn thing after another, but the result of a common fundamental, internal dynamic."[11] Thus, for example, the NATO of 1996 is very different from the NATO of 1949 and doubtless will be very different from the NATO of 2006, but its emergent properties have not transformed it into an entirely new organization. Rather, its internal dynamic has allowed it to adapt to change even though it is still in fundamental respects the North Atlantic Treaty Organization.

Adaptation and Co-evolution

But there is no magic in the processes whereby systems self-organize and develop emergent properties.

In the case of human systems, it is presumed they are composed of learning entities,[12] with the result that the dynamics of emergence are steered, so to speak, by a capacity for adaptation, by the ability of complex systems to keep their essential structures within acceptable limits (or, in the case of nonhuman organisms, within physiological limits).[13] Human systems face challenges from within or without, and the adaptive task is to maintain an acceptable balance between their internal needs and the external demands.[14] At the same time, in the process of changing as they adapt, systems co-evolve with their environments. Neither can evolve in response to change without corresponding adjustments on the part of the other. On the other hand, if a system is unable to adjust to its environment's evolutionary dynamics and thus fails to adapt, it collapses into the environment and becomes extinct. To return to the NATO example, the Organization managed from its inception to co-evolve with the Cold War and post-Cold War environments despite internal developments such as the 1967 defection of France from the military command and external developments such as the demise of the Soviet Union and the superpower rivalry. Indeed, as the environment evolved subsequent to the end of the Cold War, NATO accepted France's decision to rejoin the military command in 1996. The adaptation of NATO stands in sharp contrast to its Cold War rival, the Warsaw Pact. It could not co-evolve with the international environment and failed to adapt; in effect, it collapsed into the environment so fully that its recurrent patterns are no longer discernible.

As the history of France in NATO suggests, the co-
evolution of systems and their environments is not a
straight-line progression. As systems and their environments become ever more complex, feedback loops
proliferate and nonlinear dynamics intensify, with the
result that it is not necessarily evident how any system evolves from one stage to another. While "no one
doubts that a nation-state is more complex than a for-
aging band," and while the evolution from the latter to
the former may include tribal, city-state, and other intermediate forms, the processes of evolution do not
follow neat and logical steps.[15] Systems are unalike
and thus subject to local variations as well as diverse
trajectories through time. Equally important, evolution may not occur continuously or evenly. Even the
most complex system can maintain long equilibrium
before undergoing new adaptive transformations, or
what complexity theorists call "phase transitions." Put
differently, their progression through time can pass
through periods of stasis or extremely slow, infinitesi-
mal changes before lurching into a phase transition,
thereby tracing a temporal path referred to as "punctuated equilibrium."

The Power of Small Events

It follows from the vulnerability of complex adaptive
systems to punctuations of their equilibrium and tu-
multuous phase transitions that small, seemingly minor
events can give rise to large outcomes, that systems
are sensitive at any moment in time to the conditions
prevailing at that moment and can thus initiate processes of change that are substantial and dramatic.

Examples of this so-called "butterfly effect" abound. Perhaps the most obvious concerns the way in which an assassination in 1914 triggered the onset of World War I, but numerous other, more recent illustrations can readily be cited. It is not difficult to reason, for instance, that the end of the Cold War began with the election of a Polish Pope more than a decade earlier, just as the release of Nelson Mandela from prison was arguably (and in retrospect) an event that triggered the end of apartheid in South Africa.[16]

Sensitivity to Initial Conditions

Closely related to the power of small events is the premise that even the slightest change in initial conditions can lead to very different outcomes for a complex adaptive system. This premise can be readily grasped in the case of human systems when it is appreciated that the processes of emergence pass through a number of irreversible choice points that lead down diverse paths and, thus, to diverse outcomes. This is not to imply, however, that changes in initial conditions necessarily result in unwanted outcomes. As the foregoing examples demonstrate, the power of an altered initial condition can lead to desirable as well as noxious results, an insight that highlights the wisdom of paying close attention to detail in the policy-making process.

The Limits of Complexity Theory

Can complexity theory anticipate precisely how a complex adaptive system in world affairs will organize itself and what trajectory its emergence will follow? Can the theory trace exactly how the system will adapt or how it and its environment will co-evolve? Can the theory specify what initial conditions will lead to what large outcomes? No, it cannot perform any of these tasks. Indeed, it cannot even anticipate whether a large outcome will occur or, if it does, the range within which it might fall. Through computer simulations, for example, it has been shown that even the slightest change in an initial condition can result in an enormous deviation from what would have been the outcome in the absence of the change. Two simulations of the solar system are illustrative:

Both simulations used the same mathematical model on the same computer. Both sought to predict the position of the planets some 850,000,000 years in the future. The first and second simulation differed only in that the second simulation moved the starting position of each planet 0.5 millimeters. With such a small change in the initial conditions, [it is reasonable] to expect that the simulations would yield almost identical outcomes.

For all but one of the planets this is exactly what happened. Pluto, however, responded differently. The position of Pluto in the second simulation differed from its position in the first by 4 billion miles. Pluto's resting position is, in this mathematical model, extremely sensitive to the initial conditions.[17]

Applying these results metaphorically to the global system of concern here, it could well be presumed that the Pluto outcome is the prototype in world politics, that numerous communities and societies could deviate often from their expected trajectories by the political equivalent of 4 billion miles. The variables comprising human systems at all levels of organizations are so multitudinous, and so susceptible to wide variations when their values shift, that anticipating the movement of planets through space is easy compared to charting the evolution of human systems through time.

In short, there are strict limits within which theorizing based on the premises of complexity theory must be confined. It cannot presently—and is unlikely ever to—provide a method for predicting particular events and specifying the exact shape and nature of developments in the future. As one observer notes, it is a theory "meant for thought experiments rather than for emulation of real systems."[18]

Consequently, it is when our panacean impulses turn us toward complexity theory for guidance in the framing of exact predictions that the policy payoffs are least likely to occur and our disillusionment is most likely to intensify. For the strides that complexity theorists have made with their mathematical models and computer simulations are still a long way from amounting to a

science that can be relied upon for precision in charting the course of human affairs that lies ahead.
Although their work has demonstrated the existence
of an underlying order, it has also called attention to a
variety of ways in which the complexity of that order
can collapse into pervasive disorder. Put differently,
while human affairs have both linear and nonlinear
dimensions, and while there is a range of conditions
in which the latter dimensions are inoperative or "well
behaved,"[19] it is not known when or where the nonlinear dimensions will appear and trigger inexplicable
feedback mechanisms. Such unknowns lead complexity theorists to be as interested in patterns of
disorder as those of order, an orientation that is quite
contrary to the concerns of policy makers.

Theorizing Within the Limits

To acknowledge the limits of complexity theory, however, is not to assert that it is of no value for policy
makers and academics charged with comprehending
world affairs. Far from it: if the search for panaceas is
abandoned and replaced with a nuanced approach, it
quickly becomes clear that the underlying premises
of complexity theory have a great deal to offer as a
perspective or world view with which to assess and
anticipate the course of events. Perhaps most notably, they challenge prevailing assumptions in both the
academic and policy-making communities that political, economic, and social relationships adhere to
patterns traced by linear regressions. Complexity
theory asserts that it is not the case, as all too many

officials and analysts presume, that "we can get a value for the whole by adding up the values of its parts."[20] In the words of one analyst,

> Look out the nearest window. Is there any straight line out there that wasn't man-made? I've been asking the same question of student and professional groups for several years now, and the most common answer is a grin. Occasionally a philosophical person will comment that even the lines that look like straight lines are not straight lines if we look at them through a microscope. But even if we ignore that level of analysis, we are still stuck with the inevitable observation that natural structures are, at their core, nonlinear. If [this] is true, why do social scientists insist on describing human events as if all the rules that make those events occur are based on straight lines?[21]

A complexity perspective acknowledges the nonlinearity of both natural and human systems. It posits human systems as constantly learning, reacting, adapting, and changing even as they persist, as sustaining continuity and change simultaneously. It is a perspective that embraces non-equilibrium existence. Stated more generally, it is a mental set, a cast of mind that does not specify particular outcomes or solutions but that offers guidelines and lever points that analysts and policy makers alike can employ to more clearly assess the specific problems they seek to comprehend or resolve. Furthermore, the complexity perspective does not neglect the role of history even though it rejects the notion that a single cause has a single effect. Rather, focusing as it does on initial conditions and the paths that they chart for systems, complexity treats the historical context of situations as crucial to comprehension.

The first obstacle to adopting a complexity perspective is to recognize that inevitably we operate with some kind of theory. It is sheer myth to believe that we need merely observe the circumstances of a situation in order to understand them. Facts do not speak for themselves; observers give them voice by sorting out those that are relevant from those that are irrelevant and, in so doing, they bring a theoretical perspective to bear. Whether it be realism, liberalism, or pragmatism, analysts and policy makers alike must have some theoretical orientation if they are to know anything. Theory provides guidelines; it sensitizes observers to alternative possibilities; it highlights where levers might be pulled and influence wielded; it links ends to means and strategies to resources; and perhaps most of all, it infuses context and pattern into a welter of seemingly disarrayed and unrelated phenomena.

It follows that the inability of complexity theory to make specific predictions is not a serious drawback. Understanding and not prediction is the task of theory. It provides a basis for grasping and anticipating the general patterns within which specific events occur. The weather offers a good example. It cannot be precisely predicted at any moment in time, but there are building blocks—fronts, highs and lows, jet streams, and so on—and our overall understanding of changes in weather has been much advanced by theory based on these building blocks. . . . We understand the larger patterns and (many of) their causes, though the detailed trajectory through the space of weather possibilities is perpetually novel. As a result, we can do far better than the old standby: predict that "tomorrow's weather will be like today's" and you stand a 60 percent probability of being correct. A relevant theory for [complex adaptive systems] should do at least as well.[22]

Given the necessity of proceeding from a theoretical standpoint, it ought not be difficult to adopt a complexity perspective. Indeed, most of us have in subtle ways already done so. Even if political analysts are not—as I am not—tooled up in computer science and mathematics, the premises of complexity theory and the strides in comprehension they have facilitated are not difficult to grasp. Despite our conceptual insufficiencies, we are not helpless in the face of mounting complexity. Indeed, as the consequences of turbulent change have become more pervasive, so have observers of the global scene become increasingly wiser about the ways of the world and, to a large degree, we have become, each of us in our own way, complexity theorists. Not only are we getting accustomed to a fragmegrative world view that accepts contradictions, anomalies, and dialectic processes, but we have also learned that situations are multiply caused, that unintended consequences can accompany those that are intended, that seemingly stable situations can topple under the weight of cumulated grievances, that some situations are ripe for accidents waiting to happen, that expectations can be self-fulfilling, that organizational decisions are driven as much by informal as formal rules, that feedback loops can redirect the course of events, and so on through an extensive list of understandings that appear so commonplace as to obscure their origins in the social sciences only a few decades ago.[23] Indeed, we now take for granted that learning occurs in social systems, that systems in crisis are vulnerable to sharp turns of directions precipitated by seemingly trivial incidents, that the difference between times one and two in any situation can often be ascribed to adaptive processes, that the surface appearance of societal tranquillity can mask underlying problems, and that "other things being equal" can be a treacherous phrase if it encourages us to ignore glaring exceptions. In short, we now know that history is not one damn thing after another so much as it is many damn things simultaneously.

And if we ever slip in our understanding of these subtle lessons, if we ever unknowingly revert to simplistic formulations, complexity theory serves to remind us there are no panaceas. It tells us that there are limits to how much we can comprehend of the complexity that pervades world affairs, that we have to learn to become comfortable living and acting under conditions of uncertainty.

The relevance of this accumulated wisdom—this implicit complexity perspective—can be readily illustrated. It enables us to grasp how an accidental drowning in Hong Kong intensified demonstrations against China, how the opening of a tunnel in Jerusalem could give rise to a major conflagration, how the death of four young girls can foster a "dark and brooding" mood in Brussels, how an "October surprise" might impact strongly on an American presidential election, or how social security funds will be exhausted early in the next century unless corrective policies are adopted—to cite three recent events and two long-standing maxims.[24] We know, too that while the social security example is different from the others—in that it is founded on a linear projection of demographic change while the other examples involve nonlinear feedback loops—the world is comprised of linear as well as nonlinear dynamics and that this distinction is central to the kind of analysis we undertake.

In other words, while it is understandable that we are vulnerable to the appeal of panaceas, this need not be the case. Our analytic capacities and concepts are not so far removed from complexity theorists that we need be in awe of their accomplishments or be ready to emulate their methods. Few of us have the skills or resources to undertake sophisticated computer simulations—and that may even be an advantage, as greater technical skills might lead us to dismiss complexity theory as inapplicable—but as a philosophical perspective complexity theory is not out of our reach. None of its premises and concepts are alien to our analytic habits. They sum to a perspective that is consistent with our own and with the transformations that appear to be taking the world into unfamiliar realms. Hence, through its explication, the complexity perspective can serve as a guide both to comprehending a fragmegrated world and theorizing within its limits.

End Notes

1. A paper presented at the Conference on Complexity, Global Politics, and National Security, sponsored by the National Defense University and the RAND Corporation (Washington, D.C., November 13, 1996). I am grateful to Matthew Hoffmann, David Johnson, and Hongying Wang for their helpful reactions to earlier drafts.

2. Development of the fragmegration approach has occurred in fits and starts. See James N. Rosenau, "'Fragmegrative' Challenges to National Security," in Terry L. Heyns (ed.), *Understanding U.S. Strategy: A Reader* (Washington, D.C.: National Defense University, 1983), pp. 65-82; James N. Rosenau, "Distant Proximities: The Dynamics and Dialectics of Globalization," in Bjorn Hettne (ed.), *International Political Economy: Understanding Global Disorder* (London: Zed Books, 1995), pp. 46-64; and James N. Rosenau, *Along the Domestic-Foreign Frontier: Exploring Governance in a Turbulent World* (Cambridge: Cambridge University Press, forthcoming), Chap. 6.

3. For a discussion of the nature of these diverse "scapes," see Arjun Appadurai, "Disjuncture and Difference in the Global Cultural Economy," *Public Culture*, Vol. 2 (1990), pp. 1-23.

4. See, for example, John L. Gaddis, "International Relations Theory and the End of the Cold War," *International Security*, Vol. 17 (Winter 1992/93), pp. 5-58.

5. Cf. Roger Lewin, *Complexity: Life at the Edge of Chaos* (New York: Macmillan Publishing Co., 1992), and M. Mitchell Waldrop, *Complexity: The Emerging Science at the Edge of Order and Chaos* (New York: Simon and Schuster, 1992).

6. John H. Holland, *Hidden Order: How Adaptation Builds Complexity* (Reading, MA: Addison-Wesley, 1995).

7. Stuart Kauffman, *At Home in the Universe: The Search for Laws of Self-Organization and Complexity* (New York: Oxford University Press, 1995).

8. For a title pointing in the opposite direction, see Kevin Kelly, *Out of Control: The New Biology of Machines, Social Systems, and the Economic World* (New York: Addison-Wesley, 1994).

9. James N. Rosenau, *Turbulence in World Politics: A Theory of Change and Continuity* (Princeton: Princeton University Press, 1990).

10. As one complexity theorist put it, referring to self-organization as a natural property of complexgenetic systems, "There is 'order for free' out there." Stuart Kauffman, quoted in Lewin, *Complexity*, p. 25.

11. Lewin, *Complexity*, p. 192.

12. Holland, *Hidden Order*, p. 93.

13. The notion of physiological constraints setting adaptive limits is developed in W. Ross Ashby, *Design for a Brain* (New York: John Wiley, 1960, 2nd ed.), p. 58, whereas the substitution of acceptable limits in the case of human sys-

tems is developed in James N. Rosenau, *The Study of Political Adaptation* (London: Frances Pinter Publishers, 1981), pp. 31-40.

14. For a full elaboration of this conception of adaptation, see Rosenau, *The Study of Political Adaptation*, Chap. 4.

15. Lewin, *Complexity*, p. 19.

16. For an extensive account that traces the end of apartheid back to Mandela's links to South African President F.W. de Klerk while he was still in prison, see Allister Sparks, "The Secret Revolution," *The New Yorker*, April 11, 1994, pp. 56-78.

17. R. David Smith, "The Inapplicability Principle: What Chaos Means for Social Science," *Behavioral Science*, Vol. 40 (1995), p. 22.

18. Holland, *Hidden Order*, p. 98.

19. For the use of this phrase, see Smith, "The Inapplicability Principle," p. 30.

20. Holland, *Hidden Order*, p. 15.

21. Stephen Guastello, *Chaos, Catastrophe, and Human Affairs: Application of Nonlinear Dynamics to Work, Organizations, and Social Evolution* (Mahwah, NJ: Lawrence Erlbaum Associates, 1995), p. 1.

22. Holland, *Hidden Order*, p. 168.

23. For an eye-opening sense of how rapidly the social sciences have advanced in recent years, consider that it was only some five decades ago that, for the first time, a

gifted analyst arrested systematic attention to the dynamics of informal patterns of organizations, an insight that is today taken for granted. Cf. Herbert A. Simon, *Administrative Behavior: A Study of Decision-Making Behavior in Administrative Organization* (New York: The Macmillan Co., 1945).

24. Accounts of the events can be found in Edward A. Gargan, "Man Dies During Protest over Asian Islets," *New York Times*, September 27, 1996, p. A8; Joel Greenberg, "Dashed Hope Fed Arab Fury Against Remaining Strictures," *New York Times*, September 27, 1996, p. A1; and Marlies Simons, "Scandals Force the 2 Belgiums to Explore Inner Ills," *New York Times*, October 10, 1996, p. A3.

Complexity, Chaos, and National Security Policy: Metaphors or Tools?

Alvin M. Saperstein

Introduction

Interactions between traditional nation-states, including the extreme interaction of war, can be likened to the interaction between microscopic bodies in physics. Relatively few variables are required to describe the process; the course of events is basically predictable—between occasional major, contingency-based, bifurcations (e.g., the outcomes of specific battles or collisions). Subnational wars—ethnic or tribal conflicts, guerrilla insurgencies—would then have to be likened to the interactions of meso-physics: fluctuations away from the mean become at least as important as the mean. The descriptive words usually resorted to are chaos, complexity, non-pre-dictability, etc.

In the modern era, the actual and potential destruc-tiveness of inter-nation war has tended to stabilize S.U. (Soviet Union)-U.S. type conflicts—with their nuclear weapon implications. This has allowed the realm of ethnic type war possibilities to grow and with it the attention of policy makers, scholars, and soldiers to the concepts of chaos and complexity—the theme of this conference.

That the paradigm of chaos was intimately associ-ated with battle was certainly well known to von Clausewitz and the earlier Greek military historians. Many of the people at this conference, whose writ-ings I have read with pleasure and profit (Beyerchen 1992, Lane and Maxfield 1996, Mann 1992, Mazarr 1994, Rosenau 1996, Rinaldi 1995, Schmitt 1995), have made amply clear the usefulness of the com-plexity concept in describing international security strategy.[1] But do we gain anything from the visits of the soldier and statesman to the academy of the math-ematician and physicist, besides some new, exotic descriptive metaphors (e.g., "strange attractor," "self-organizing criticality")?

Do we gain any useful policy making and/or strategic tools as a result of the concordance of the new meta-phors, derived from the physical sciences, with the long recognized *chaotic-complex* aspects of war and national security in a competitive anarchic world?[2] Has anything been gained by the transfer of the growing popularity of these paradigms from "hard" to "soft" scientists or the recognition of the growing prevalence of these "fads" by the military and political elites? A new set of metaphors to describe a world does not imply new or different behaviors of that world—*we must be very careful not to confuse changes in an intellectual outlook with changes in world events or patterns which we hope to <u>understand and master</u>.*

The role of the policy maker, whether in a domestic or an international system, is to <u>master</u> the system: to be able to take actions now which will lead to desirable events, or avoid undesirable events, in the future. Thus he/she must be able to <u>predict</u> the outcome of current activities: if I do <u>A</u>, <u>A</u>' will result; if I do <u>B</u>, <u>B</u>' will result, etc. Prediction is the transfer of knowledge of a system from its present to its future. The ability to make such transfers is usually based upon an understanding of the system—unless recourse is made to auguries or direct communications from a transcendental power. Excluding the roles of divination or divinity, we must help the rational policy maker to <u>understand in order to master</u>.

It is clear that the set of metaphors which underline our thoughts and discussions about the political world determine our responses to matters of war and peace.[3] Action often follows theory. (But purely pragmatic responses—not the best, but adequate—are often resorted to by some societies with some success. Non-theoretical societies do survive, sometimes.)

Moreover, we also recognize that our metaphors may also shape that political world.[4] The "field of endeavor," within which we are trying to find appropriate responses, is not itself fixed apriori; its contours may be molded by our metaphors; the topographic maps relied upon by the competing forces may be altered by the plans and actions of these forces. Hence policy and response are easier and more effective, the more appropriate the available metaphors.

It should also be clear that the new metaphors will be helpful in educating that majority of citizens, soldiers, and statesmen which have not experienced chaos and complexity due to the apparent simplicity of the bipolar world view of the last half-century. It may be easier to have university freshman and military cadets read modern works on complexity and chaos (e.g., Gleick 1987, Waldrop 1992) than have them study Thucydides or von Clausewitz. Metaphors also determine the social acceptability of presenting ideas publicly, thus subjecting them to criticism and possible action. For example, without the intellectual possibility of the dissolution of nations, i.e., complexity, few conceived of (and thus planned for) the end of the Soviet Union (and even fewer for that of its Cold War partner, the U.S.). The new intellectual paradigms should focus attention on the underlying world political realities—chaos and complexities which have always been there, sometimes obscured to many, but always recognized by some.

It is important to recognize that our metaphors, just as our goals, the "fields of competition and endeavor," and the events themselves, are constantly changing as a result of our formulating ideas, exploring our world, and attempting to control events and reach goals. We must be careful not to imbed our ideas and "world-pictures" in stone since the stone of the world is often brittle and ruptures catastrophically, or flows and deforms like lava. "He that will not apply new remedies must expect new evils, for time is the greatest innovator." (Philosopher-statesman Francis Bacon, 17th century)

Metaphors—Old and New

There are two major classes of metaphors, with roots in the history of physics, that are appropriate to this conference on global politics and national security: The Newtonian view is that of a fixed set of elements. They interact, linearly or non-linearly, in a fixed universe. Depending upon the issue under discussion, these elements (and their interactions) may be: nations interacting with each other (via war, negotiation, trade, cultural or terrorist exchange,...) in a world system; economic, bureaucratic, class,... groups "pressuring" each other within a given nation; military divisions, regiments, battalions,..., engaged with each other in battle or along a front; etc. The strengths of the individual elements and of their interactions may wax or wane, their "location" in the "field of endeavor" may change with time, but their continued existence, as well as that of the system of which they are elements, is taken for granted. (In the wars of kings, it was usually assumed that the opposing king would still be there "afterwards," just somewhat diminished.) This Newtonian paradigm of sovereign nations has been the usual framework for discussions of international security during much of the past few centuries.[5]

In the currently fashionable Prigoginean[6] (Prigogene and Stenger 1984) paradigm ("self-organizing criticality"=SOC), elements and their interactions come into and go out of existence as part of the ongoing process; the field of endeavor may change in size, structure, and constituents with time. Thus states, armies, military and civilian units, may be born, grow, thrive, decay, die and disappear, as part of the process which also creates, distorts, and dissolves, the structures of which they are—if perhaps only temporarily—parts and foundations.[7] States may be created out of, or dispersed back into, smaller groups of people as a result of war or other interactions between other states or people groupings.[8] "Official" or "unofficial" military units form or dissolve as a result of anticipated or actual conflict between existing, nascent, or hopeful nations.[9] Economic, political, or other classes, come and go through turmoil engendered by other groupings in the system of nation or nations.[10] In sum, *the system determines its apparent elements rather than conversely.*[11]

The changing of the elements, their interactions, and the overall structure may occur at vastly different time scales. Consequently, there may be intervals of time in which the system seems to consist of fixed elements interacting with each other under fixed rules, i.e., a

Newtonian description may provide a good approximation for some epochs. Conversely, a Newtonian system of small enough elements may provide the conceptual foundation for a Prigoginean system of larger elements: the shifting elements of the latter may "actually" consist of varying combinations of the fixed elements of the former. For example, guerrilla bands, regiments, tribes, nations, states, are all different time-varying combinations of people; the underlying Newtonian system would be the multi-billion member set of the world's population. (And, of course, each person is a shifting combination of biological cells. And, each cell is a shifting sets of molecules. And so forth.)

Both of these paradigms can be taken with either a stochastic or a deterministic view. In a stochastic model there are no rules connecting the state of the system at one instant of time deterministically to its state at a following instant. Only probabilities connect the two. Within a stochastic Newtonian model, interactions between elements can be likened to the random collisions of molecules. Policy can be framed by comparing the relative probabilities of the outcomes of different policy-choice-paths and maximizing expectation values. Combining the stochastic and Prigoginean metaphors, security interactions would be modeled by "collisions" between elements which may or may not exist. Without resorting to the full apparatus of quantum field theory, there is no obvious simple means of rationally dealing with such models, and so they will be avoided in this paper.

Deterministic systems have rules, which may be ascertained, which uniquely connect neighboring time states of the system (Fig. 1a). In Newtonian systems, these rules would govern the interactions between the permanent elements. Within the Prigoginean paradigm, the rules would also govern the creation and dissolution of these, now perhaps impermanent, elements. Most people act, and have acted historically, as if there are "rules of human behavior." Hence I will stick to deterministic paradigms.

It is important to stress that <u>determinism</u> does not imply <u>predictability</u>. Prediction implies connections of necessity (not of probability!) between non-perfectly well-defined states of the system separated by finite time intervals. In order to rationally predict future behaviors of a system, we must know its present state. If the future knowledge so obtained is roughly comparable in quality to the present knowledge, the prediction is successful. But present knowledge is never perfect. *There are always measurement errors in any determination of the present state.* The resultant non-perfectly well-defined <u>present</u> state encompasses a number of possible starting states. The rules determining <u>future</u> states must be applied to each of these starting states. Thus, given any deterministic model, implicit or explicit, upon which predictions are to be based, a range of "paths into the future" are possible (see Fig. 1b,1c). Furthermore, any such model depends upon *parameters obtained from necessarily imperfect observations.* Hence even

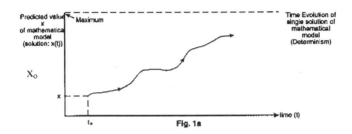

Fig. 1a

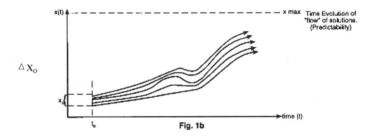

Fig. 1b

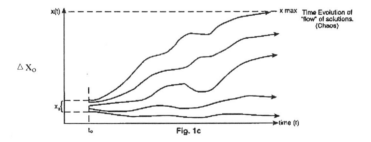

Fig. 1c

a perfectly determined initial state of the system allows a range of future outcomes in any reasonable predictive modeling.

The result of these two imperfections of observation is that any set of rationally ascertained system rules, which transfer realistically obtained present knowledge of the system into the future, will result in a range of possible outcomes—a range of uncertainty. If this future range of uncertainty is large compared to the range of present knowledge, the quality of prediction is impaired. If this future range covers all possible outcomes of the system (Fig. 1c), no knowledge of the future is possible—prediction (and hence rational policy making) is impossible.

If the rules governing the system are "linear,"[12] the range of future outcomes is always comparable to the range of input uncertainties (Fig. 1b): prediction is possible, and therefore useful to the policy maker. If the system rules are non-linear[13] (as are most systems involving competing human beings, wherein the policy of one party must not only include the desired goals of each party but also the response of the other parties' progress toward those goals[14]), the system may display *extreme sensitivity* to small changes in input or system parameters (Fig. 1c). This behavior, called "chaos," (see, e.g., Schuster 1988) makes prediction—and hence control of future behavior of the system—difficult or impossible. However, it may be possible to predict whether or not a system will display chaotic behavior. This possibility, as shown in the following section, allows the policy maker to avoid dangerous behavior. Hence *the ability to predict unpredictability is a very useful tool in policy making* (Saperstein 1986).

Crisis Instability and Chaos

In pre-WWI Europe, the assassination of two people in the Balkans was enough to ignite a carnage that swept all of the continent and involved all other continents, left millions dead, vast territories desolate, wiped out existing nations and governments, and created new ones. In post WWII Europe, the murder of hundreds, or perhaps thousands, in these same Balkans left most of the rest of the world untouched— except perhaps in their consciences and charitable purses. In the first case, a very small change in the system parameters led to major transformations of the system—the definition of "chaos" if the system were a mathematical/physical system. In the second case, the disturbances effectively damped out as they propagated through the system—the sign of a stable mathematical/physical system. The political scientists have coined the phrase "crisis instability" to describe the first case—extreme sensitivity of the world political system to minor perturbations (see, e.g., Saperstein 1994). In the second case, the world system was "crisis stable." The same world system can manifest crisis instability at some places during some epochs, while displaying crisis stability elsewhere or at other times.

Physical systems, e.g., a moving fluid, can also display chaos (i.e., turbulence) in some circumstances and stability (i.e., smoothly varied, ordered, laminar flow) in others. Mathematical metaphors for these physical systems must be able to manifest both chaos, stability, and the transition between the two, if they are to be a reasonable representation of the physical reality. Furthermore, if it is to be useful, the mathematical model must be able to predict the circumstances under which the system will switch from stability to chaos. For example, the airflow over a given wing design will be laminar for air velocities whose Reynolds's number[15] is below some critical value (Fig. 2). For larger flow velocities, the flow becomes turbulent, dissipating energy in an uneconomical manner and making control of the total aircraft more difficult and perhaps dangerous. The ability to predict the critical Reynolds number, and its variation with changes in aircraft design, is very important for the aircraft designer who wishes to avoid having to find out that his aircraft is unstable via the sacrifice of test pilots or passengers.

Analogously, if it were reasonable to mathematically model the world system of nations, a chaotic mathematical system would be a good metaphor for a crisis—unstable world. Being able to predict the critical "Reynolds number" for such a world model would be very important for the policy maker whose goal was to avoid crisis unstable conditions with their concomitant high probabilities for the outbreak of war (Saperstein 1984).[16] (In the modern political/weapons-of mass-destruction world, there are no "test pilots" and we are all potentially sacrificial passengers.)

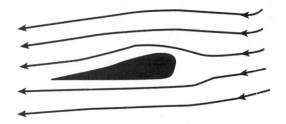

R<R$_c$:Laminar Flow

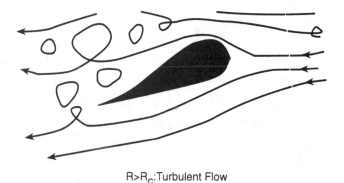

R>R$_c$:Turbulent Flow

Fig. 2

In a Newtonian world paradigm (or in a Newtonian approximation to a Prigoginean world view), the notion of national security—and the goals of the corresponding policy makers—are fairly straightforward. Policy must be framed so as to either avoid war or to reap the benefits of winning a war (whose win can be "guaranteed" with associated costs less than expected gains). In either case, the prime goal is to maintain control of the future, to retain predictability and hence avoid crisis instability. Given a reasonable mathematical model of the system for which policy is

being made, it can be used to explore for system characteristics which allow transition to chaos. The
policymaker must then studiously avoid the corresponding behaviors or conditions.

An example of interest to the strategist of bipolar
nuclear arms races (in the context of the S.U. - U.S.
Cold War) is the modeling of the Strategic Defense
Initiative, the proposal during the Reagan Presidency
to deploy a massive system of ground-based and
space-based defenses against strategic-ranged bal-
listic nuclear missiles. The model (Saperstein and
Kress 1988) presumed that each of the two antago-
nists would deploy similar offensive and defensive
systems against the other (Fig. 3). The deployment
numbers would be determined in response to the
opponent's deployed weapons numbers; the result is
a non-linear interactive system whose stability can be
investigated by conventional means: introduce a small
disturbance into the system and compute how it grows.
As expected, there are starting configuration numbers
(of offensive and defensive missiles) for which the perturbations remain small, others for which they grow
greatly and rapidly (Fig. 4). The latter configurations
are the crisis-unstable systems which are to be
avoided by the relevant strategic planners.[17]

The same paradigm has been used to explore questions of more academic interest. Using a non-linear
Richardson[18] model of the arms race between competing nations, a comparison (Saperstein 1991) was
made of the stability region of three-nation systems
(Fig.5a) with that of two-nation systems (Fig.5b). The

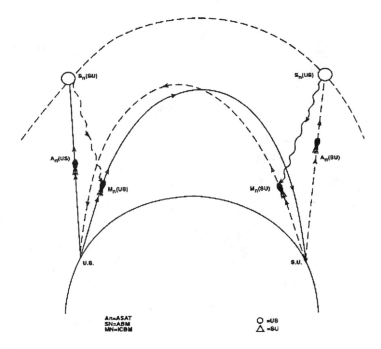

Fig. 3

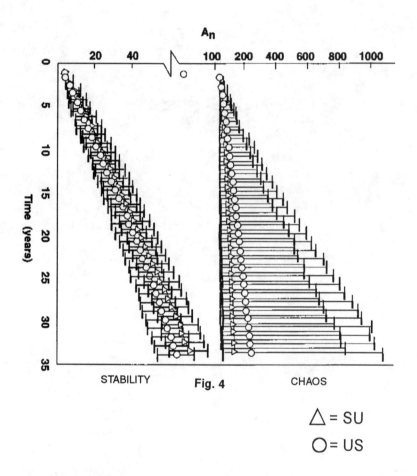

Fig. 4

STABILITY   CHAOS

$\triangle$ = SU
$\bigcirc$ = US

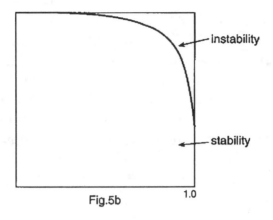

Fig.5b

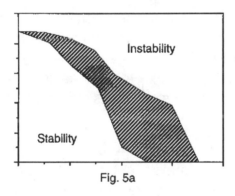

Fig. 5a

former was found to be smaller than the latter, indicating that it is more difficult to stabilize a tri-polar world than a bi-polar world, a conclusion which has also been drawn by many "conventional" non-mathematical political scientists. Another concordance between the results of mathematical modeling of international systems and conventional analysis has been that a system of democratic states is less likely to have wars than a system including oligarchic states. The model conclusions (Saperstein 1992a) result from the differ-

ing values of the Richardson-type parameters[19] stemming from democratic versus oligarchic societies. The
differences arise since the (Newtonian) nation entities of the Richardson model, and hence their
interactions, result from averages over a larger
Newtonian model whose elements are the nation's
decision makers—citizens, politicians, officials—a
large class in the democratic state, a small group in
the oligarchic state. In the latter case, the interaction
parameters resulting from the average are more likely
to be large enough to produce an unstable system.
Finally, a comparative stability analysis was made of
systems of competing nations, each looking out for its
individual security, versus systems of alliances, shifting so as to maintain a "balance of power" (Saperstein
1992b). Again, the result—that it is easier to stabilize
a balance-of-power system—was expected from conventional political analysis.

In all of the above cases, the chaos metaphor was
used to steer policy makers away from potentially dangerous crisis instability situations—away from chaos.
Alternatively, when war and its associated chaos is
unavoidable, there is the traditional approach to the
chaos of battle, an approach used by successful military planners whether or not they recognized or used
the chaos metaphor. Since small perturbations can
lead to largely different outcomes ("For want of a nail,
a shoe was lost,... a kingdom was lost.") one appropriate response (characteristic of the U.S. military since
Grant) has been to always deploy overwhelming
forces, if they can be made available. (Have more than

enough horses, so that the loss of a few would make no difference.) That is, the statistical fluctuations which mimic chaos usually scale as the square-root of N, the number of significant elements. For large enough N, the relative fluctuations are unimportant. An alternative to increasing the sizes of the force units available (the Newtonian elements of the system) is to increase the number of different types, their flexibility and rapid adaptability to changes. Have horses, mules, people, jeeps, well trained and available to carry out the required tasks. Better yet, have available alternative sets of tasks and immediate goals, which will lead to the final desired goal—if you can't take that hill, take the other one. It is clear here that the new chaos metaphor offers no new tools to the military planner though, as has been previously suggested, it may significantly aid the military educator.

National Security in an "SOC" World

The goals of the national Security policy maker are not so obvious in a Prigoginian ("Self Organizing Criticality") world. Should policy be aimed at encouraging or discouraging the creation of new nations, the breakup of the old? Should new alliances, new armies, new bones of contention be anticipated? All of these are the possible system elements and interactions (between the elements) which may arise and evolve via the life of the system. It is now clear that all of these SOC possibilities must be anticipated as well as the vagaries of dealing with the usual interactions between the Newtonian elements of long-lived nation's and

alliances. For example, should the "West" have encouraged the break-up of Yugoslavia? (There is a
long history of eastern European people living at peace
with each other in strongly ruled, multinational, nondemocratic States.) Are we better off competing with
oppressive but strong oligarchies or dealing with fragmented—even worse, fractal—democracies?[20]

One of the prime reasons for our failure to successfully deal with Iraq—a "sovereign" element in the
Newtonian system—is that we fear to deal with its possible break-up. Similarly, there were important
confusions in our society in anticipating and dealing
with the break-up of the Soviet Union. Our policies
towards China have also suffered from these confusions. In the Newtonian scheme-of-things, nations are
sovereign states and deal only with each others' sov-
ereigns. "Infringing upon sovereignty" is severely
frowned upon. It is clear that we still speak to such a
world, though we no longer live it.

It is not evident to me that a single metaphor/tool—
like chaos—is available or useful to us in dealing with
a world system characterized by "complexity."

Instead of specific new tools, these metaphors can
contribute to the development of the new attitudes required for the more complex modern world. They can
help sharpen minds dulled by a Newtonian world view
so as to be alert to all new possibilities. (It should be
obvious that such alertness and openness was always
present in some outstanding historic leaders whose
minds were, perhaps, not so overburdened with

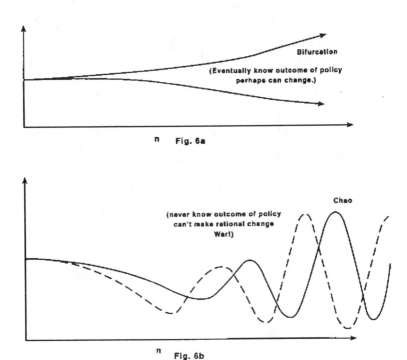

Fig. 6a

Fig. 6b

Newtonian simplicities.) Above all, we have to be alert to (and be able to respond to) the possibility of bifurcation[21] (Fig.6a) of the existing system into very different possible worlds, containing new and different elements interacting in novel ways. Such bifurcations may occur at national levels—where nations rise and fall, where they are of interest to the strategist, and at local levels—of tactical interest, where military, governmental, or corporate units are created or destroyed. Though these bifurcations are contingent, the probabilities of their occurrence, and their outcomes, are not structureless; familiarity and insight into the fundamental aspects of the system can lead to clues as to when the probabilities of such change are large, and when they are small.

Thus we shall need very flexible diplomats and soldiers at all levels.[22] (The metaphors of complexity may be helpful in recruiting as well as in educating them.) They will have to be very knowledgeable about past behavior of the system and its elements—as determining the chances for radical transformation of the system. They will have to be open and adaptable to the new and novel which may confront them - with or without rational anticipation.[23] Clearly, the new policy makers will have to be thoroughly cognizant of the relevant elements of anthropology, sociology, and psychology, as well as history. Knowledge of the functioning of existing governments, their departments or military units, will not be sufficient, as these elements may be bubbling-up or dissolving into the inchoate foam of people and groups below.

Not only are flexibility and imagination required for attaining one's ends in a complex system. The ends themselves will often be shifting and/or unclear. In some cases it may be desirable to fragment competing parties ("divide and conquer"—e.g., the British role in India); in other cases to consolidate them (create alliances or nations—e.g., the creation of Yugoslavia[24]). Of utmost importance is the recognition that the policymaker can help direct these shifts, by influencing the elements at a lower level than those of the system of interest; e.g., in a system of nations, it may be advisable to attempt to influence their individual citizens.[25] So much for the sanctity of national sovereignty!

In mathematical terms, the usual way of seeking the "best" solution to a problem is to look for some maximum value of a function-surface over the space of values pertinent to the problem (e.g., Axelrod and Bennett 1991). The highest maximum (or the lowest minimum) is the best solution—the desired policy— and if the surface is known, that best solution can eventually be found. However, in a "Self-Organizing Criticality" world, the act of moving over the surface in search of its maximum can radically change the surface. It will thus act more as an elastic membrane than as a fixed-function surface. Thus we may not be able to look for the "good strategy" in opposition to the "bad strategy" but may have to settle for the "contextually appropriate strategy."

Conclusion: Chaos and Complexity— Tool and/or Metaphor?

It is clear that successful military and political policy makers have always entertained the potentiality of chaos and have sought the tools of redundancy and flexibility of resources to deal with that possibility. The only new tool to deal with chaos presented here is the engineering tool of attempting to predict crisis instability and then avoid it or be prepared to live with it. Quantitative dynamical models of the system of interest may be useful in making such predictions. If they are inadequate or unavailable, verbal models have a long history, and potentiality, of use.

If the leaders of the pre-WWI European states had recognized that the railroad schedule-dominated mobilization of their troops was a source of great crisis instability (Tuchman 1962, van Creveld 1989), perhaps they would have avoided starting—and being trapped by—the process. But this recognition would have required that the chaos metaphor be more commonly found in the "intellectual air" of turn-of-the-century Europe than was the case in that rapidly industrializing Newtonian-reductionist society.

Given a Newtonian paradigm, the policymaker strives to be efficient in reacting to a given "field of endeavor"; chaos is to be avoided or dealt with by overwhelming force and/or redundant means of force delivery. The present world seems to require a Prigoginean outlook: don't accept the battlefield or the world system as a fixed given. The complexity, or adaptive self-organizing, metaphor should be very useful for the necessary education, recruitment, planning, and thinking required to deal with and survive our future. However, no obvious specific tool—like predicting crisis instability—comes to mind. The metaphor require that one should always be contemplating the future. And, among these considerations for the future, always include attempts to change the field of endeavor itself.

Hence, it may not be useful for the policymaker to always look for the uniquely "best solution." It may be necessary to settle for a local temporary maximum— a good solution, rather than the best. In the elastic fabric of our present and future world, the "perfect" is often the enemy of the "good."

When all is said and done, on a strategic level, the most useful aspect of the chaos and complexity metaphors is to remind us and help us to avoid falling into chaos.[26]

End Notes

1. "Complexity may be defined as the set of deterministic theories that do not necessarily lead to long-term prediction....The numerical variables are still uniquely related to each other locally in space and time. But...we cannot obtain the future values implied by the theory just as a result of compact, well-defined manipulation of the present values....Complexity theories thus depend on the complete 'path' taken by the system between its beginning and end points....Every intermediate instant of time may see the theoretical system diverted from the path it might have taken in the absence of perturbations, which are always present....The system is extremely context-dependent." (Saperstein, 1995)

2. Contrary to popular wisdom, it may not be so bad to be prepared to fight the last war! Last wars have always been chaotic and complex; it is only in the post World War II "cold war" that some *serious* stategists have believed in a non-complex world paradigm.

3. A "non-appeasement" world view, stemming from the failure of appeasement towards Hitler, has governed our post-WWII policies towards Stalin, Iraq, Bosnia...

4. For example, the Wilsonian ethnic metaphor—that every ethnic nation should have it own state—broke up the European multi-ethnic empires, leading (?) eventually to disasters like the Bosnian conflict.

5. "Do what you wish to your own people and your neighbors will not get involved." "Zaire Under the Gun," *New York Times*, Nov. 3, 1996, p.E3.

6. I am indebted to John F. Schmitt (1995) for this characterization.

7. In a complex adaptive system, these "emergent properties" or "structures" are the result of contingency, not determinism: you cannot *predict* when, or if, they will emerge, how long they will endure.

8. Zaire is a national state now—but for how long?

9. From whence did the Taliban militas come; will they last?

10. The Russian "Mafia" may be such an "emergent" "business class."

11. In the perpetual intellectual dispute between "wholeness" and "reductionism" (the whole is different from the sum of its parts vs the whole is equal to the sum of its parts), SOC is in the wholeness camp.

12. Changes in output are proportional to changes in input; equivalently, the output resulting from the sum of two inputs is equal to the sum of the two outputs separately resulting from each of the two inputs.

13. Non-linearity implies that the anticipated response to a planned action modifies the plan.

14. As an example of non-linear behavior, consider a nation, pacific in intent, which only arms itself in anticipation of possible attacks from its explicitly aggressive neighbor. It realizes that the neighboring nation will detect its arms buildup and respond with its own; in fact the neighbor might be inclined to advance its presumably planned attack so as to come in ahead of the determinedly defensive arms buildup.   So, in anticipation of this response, the defensively oriented nation launches a supposedly preemptive attack against the presumed aggressor!

15. A dimensionless system parameter which is determined by the characteristic size and flow velocity as well as by the viscosity and density of the fluid and which determines the properties of the fluid flow. When the Reynold number excedes the critical value (determined by the basic characteristics of the system) the system becomes unstable to transition to a chaotic state.

16. Certainly, close thoughtful attention to the developed world's hungry reliance upon petroleum, imported from regions controlled by closed oligarchies, should have raised the prospect of impending crisis instability.

17. This warning of the possibility of a loss of predictability and control over an escalating arms race came at a time when some optimistic Cold-War strategists were arguing for the practice of *precise control* over an upward spiraling MAD dance.

18. The usual linear Richardson model of a two-party arms race assumes that the rate of acquisition of arms, by each party, is proportional to ("linear in") the existing stock of arms of its opponent and to its own arms stocks. The non-

linear model takes into account the possibility that the opponent's stocks can become "saturated" and hence of diminished danger.

19. The coefficients of proportionality between the existing arms stocks and the acquisition rates for new arms—hence a measure of the distrust and fear of the opponent and the confidence in one's own arms.

20. Czechoslovakia fragmented into the Czech Republic and Slovakia. Unfortunately for the people of Bosnia, the different ethnic groups living there have fractal boundaries between them. In the former case, there are clearly two separate areas, separated by a reasonably "smooth" boundary; this is not true in the latter case.

21. Bifurcation (Fig. 6a) represents a choosing (in the usual way) one of several possible futures (which contingently become available), leading to the creation of sets of distinct plans—one for each future. Chaos (Fig. 6b) implies that these different futures are interbraided. Hence plans must constantly be mixed and revised.

22. In a chaotic situation, every element must be prepared to become a Clausewitzean "center of gravity" if the designated center is knocked out. The German tanks did so well early in WWII, against their technologically equal or superior opponents, because each one was equipped with radio and each understood the goals and rationale of the original plans and so was able to take over and modify plans as necessary.

23. A good football team may have separate offensive and defensive squads, but each must be able to fulfill the role

of the other when circumstances (fumble, interception) so require—which is often. In the military, it may be possible to make do with a previously designated and trained "peace-keeping quarterback" and a "peace-making quarterback," etc., each prepared to take over and lead a well trained "general-purpose squad" for the appropriate purposes. We know and expect that ordinary military units can carry out diverse tasks.

24. Note that the same "world system" sometimes finds it useful to consolidate, and sometimes useful to fragment its previous consolidation, e.g., Yugoslavia.

25. Such influence has long been attempted, e.g., Voice of America, BBC Overseas Service, "hidden" subsidies to the political parties, labor unions, business enterprises, newspapers, radio, TV, etc., of other countries, and of course, propaganda to troops on and behind the front lines.

26. The author is greatly indebted to his colleague (and wife) Harriet for her careful reading of the first draft leading to critical, insightful, and productive suggestions.

References

Axelrod, Robert and Scott Bennett. 1991. *A Landscape Theory of Aggregation.* University of Michigan preprint.

Bacon, Francis. 17th century. Quoted in *Newsweek*, Oct. 14, 1996, p.43.

Beyerchen, Alan. 1992. "Clausewitz, Nonlinearity, and the Unpredictability of War." *International Security* 17:3, 59-90.

Gleick, James. 1987. *Chaos: The Making of a New Science.* New York: Viking.

Lane, David and Robert Maxfield. 1996. "Strategy Under Complexity: Fostering Generative Relationships." *Long Range Planning,* 29:2, 215-231.

Mann, Steven R. 1992. "Chaos Theory and Strategic Thought." *Parameters* Autumn, 54-68.

Mazarr, Michael J. 1994. *The Revolution in Military Affairs: A Framework for Defense Planning.* Carlisle Barracks, PA: U.S. Army War College.

Prigogen, I. and I. Stengers. 1984. *Order Out of Chaos.* New York: Bantam.

Rinaldi, Steven M. 1995. *Beyond the Industrial Web: Economic Synergies and Targeting Methodologies.* Maxwell Airforce Base, School of Advance Airpower Studies, Air University.

Rosenau, James N. 1996. *Complex Humanitarian Emergencies: Toward an Integrated Understanding.* Cambridge: Center for Population and Development Studies, Harvard University. April 1.

Saperstein, Alvin M. 1984. "Chaos - A Model for the Outbreak of War." *Nature* 309, 303-5.

Saperstein, Alvin M. 1986. "Predictability, Chaos, and the Transition to War." *Bull. Peace Proposals,* 17:1, 87-93.

Saperstein, Alvin M. and Gottfried Mayer-Kress. 1988. "A Non-linear Model of the Impact of S.D.I. on the Arms Race." *Journal of Conflict Resolution* 32:4, 636-70.

Saperstein, Alvin M. 1991. "The Long Peace - Result of a Bi-Polar Competitive World?" *Journal of Conflict Resolution* 35:1, 68-79.

Saperstein, Alvin M. 1992a. "Are Democracies More or Less Prone to War? A Dynamical Model Approach." *Mathematical and Computer Modeling* 16:8/9, 213-221.

Saperstein, Alvin M. 1992b. "Alliance building vs Independent Action: A Non-linear Modeling Approach to Comparative International Stability." *Journal of Conflict Resolution* 36:3, 518-45.

Saperstein, Alvin M. 1994. "Chaos as a Tool for Exploring Questions of International Security." *Conflict Management and Peace Science,* 13:2, 149-77.

Saperstein, Alvin M. 1995. "War and Chaos." *American Scientist,* 83, 548-557.

Schmitt, John F. 1995. *Chaos, Complexity and War: What the New Nonlinear Dynamical Sciences May Tell Us About Armed Conflict.* Quantico, VA.: Concepts & Doctrine Division, Marine Corps Combat Development Command.

Schuster, H.G. 1988. *Deterministic Chaos—An Introduction.* Weinman, FRG: VCH Verlagsgesellschaft mbH.

Tuchman, Barbara. 1962. *The Guns of August.* New York: Dell.

van Creveld, Martin. 1989. *Technology and War.* New York: the Free Press.

Waldrop, M.M. 1992. *Complexity.* New York: Simon and Schuster.

The Reaction
to Chaos

Steven R. Mann

I want to talk about the art of foreign policy. And the art of strategy. And the art of diplomacy. And of course the art of war. Individually, these are throwaway phrases. But I think there's a deeper truth imbedded in this idea of art and policy affairs. The truth relates to the terrific need of humans to develop order. This is the mission of Western art—the Western Eye—in imposing form on nature and calling it beautiful. Art is at war with nature. It is the art of foreign policy that seeks to impose structure on the environment and build a pleasing stability. It is no accident that we refer to these different "arts." So by discussing art, I am saying that this is a talk not merely about chaos, but a practitioner's view of the ways in which we react to chaos. (Here I have drawn liberally from Camille Paglia and have extended this to the political.)

That the world is chaotic is another throwaway line. Even in the policy community, where so many of us earn our living, this is a commonplace. In actual practice, though, we the United States are wary of going beyond the commonplace in addressing the fact and implications of chaos, or better said, addressing the dynamical nature of the world. In this talk, I want to discuss further why this is difficult and suggest what implications this has for our policy direction. First, though, let me make the case that we are in a chaotic world.

The argument that I am going to make is that foreign affairs exhibits characteristics of self-organized criticality. Briefly stated, the tenets of self-organized criticality are these: "many composite systems naturally evolve to a critical state in which a minor event starts a chain reaction that can affect any number of elements in the system. Although composite systems produce more minor events than catastrophes, chain reactions of all size are an integral part of the dynamics. According to the theory, the mechanism that leads to minor events is the same one that leads to major events. Furthermore, composite systems never reach equilibrium but instead evolve from one metastable state to the next."

What first led me five years back to the metaphor of self-organized criticality was the utter unbelievability of the phrase, "the New World Order." Whatever else we were seeing in international affairs, it was not order. But that phrase has got legs; it even appears in the conference brochure. Leaving aside the unfortunate conspiratorial echoes of that phrase, which have fueled militia paranoia across the US, it is not properly descriptive. I'd suggest instead that what we are actually dealing with is better expressed by the concept of consistent criticality. The international environment is complex, dynamic, and constantly changing. The world appears as a critical arena.

The destruction of the old paradigm of an ordered, bipolar international environment meant therefore that there would be a nostalgic, sometimes obsessive drive to reclaim the idea of a stable international scene. Thus the new world order.

And indeed, something other than order is at work here. Look at the out-of-the-ordinary array of international crises that we have experienced in the past five years: Somalia, Haiti, Bosnia, Central Africa, Chechnya. This is not to speak of second-tier crises (from the US view) such as Abkhazia and Kashmir. I believe we are in an environment in which unpredictable transformations lead to constant change in the international environment—yet throughout these various upheavals, the overall system retains a surprising degree of robustness. The model of self-organized criticality wears well in describing this policy environment.

In order for events to proceed to the global critical state, there needs to be a sufficiently complex international system in existence. To achieve truly global criticality, which is what we are seeing in the Twentieth Century, the following prerequisites were necessary: efficient methods of transportation; efficient means of mass production; greater freedom in economic competition; rising economic standards, leading to greater weight on ideology [when the struggle for survival is won, there is room for ideology]; efficient mass communication; and an increase in resource demand. I'm sure this is not an exhaustive list, but these items occur to me as necessary preconditions for global criticality. Talk of global complexity is also a commonplace, usually expressed in terms of "global interdependence." But I think global criticality is a more productive way of thinking of this.

Now, you can indeed take this stuff too far. Social sciences are often subjective. Chaos theory has become trendy. It's easy to overstate the power of a theory. This brings us to the "Is it live or is it Memorex? question." Do chaos and self-organized criticality exist as actual principles of international affairs or are we dealing with perceptions and metaphors? Vice President Gore has called criticality "irresistible as a metaphor." This is true and we need to use caution. Humans have a terrific need for stability and one of the ways we serve this need is through the search for paradigms. We consider reality tamed if we find a classification, a description for it. But I no longer believe in criticality merely as metaphor. I think that the process involved is a "real" one, not a representational one. I believe that the action of international actors is an actual example of a chaotic environment and that out of the interplay of large numbers of actors with great degrees of freedom, we are seeing self-organized criticality on a global scale.

The idea of chaos and criticality operating in the social arena is becoming more and more accepted. I read of applications of chaos theory to economics.

Especially intriguing to me has been the attention given to dynamical systems theory by psychoanalysts. This strikes me as a bold and plausible application of these theories to the "soft" sciences—sciences where quantification is difficult and the dangers of subjective analysis are high—and I believe we as strategic analysts would do well to study this research. One of these psychoanalysts, Dr. Galatzer-Levy states: "Chaos theory was born out of the recognition of what could not be done." <u>What could not be done</u>. Recall the discomfort I, and I'm sure many of us had as we tried to make sense of the "New World Order." Matching chaos theory to psychoanalysis, Galatzer-Levy writes: "Any reasonably complex system is not predictable in its details over a long period of time. Certainly the human mind is at such a level of complexity." Now, if we are dealing with the product of billions of human minds in an interactive, responsive system, is it not reasonable to believe that chaos theory applies to our particular science?

Galatzer-Levy asserts that he finds in psychoanalysis such dynamical phenomena as strange attractors and self-similarity. Earlier, two other analysts, Sashin and Callahan created a model of affect—the emotional response attached to a stimulus—using catastrophe theory. We need to be open to these concepts as actual phenomena, not merely as metaphor. In our own field, we need to take encouragement from what these observers suggest and develop a consistent model of international affairs that incorporates dynamical systems theory. A successful model—if it can be created—will encompass military strategy, trade and finance, ideology, political organization, religion, ecology, mass communication, public health, and changing gender roles. For better or worse, the sum of these topics is foreign affairs today.

Twentieth-century history alone has ample evidence for the idea of criticality—though here of course we have again to be careful of subjective interpretation. This century's history exhibits a recurring pattern, of building to a critical state, catastrophic change, subsequent reordering, and a period of metastability—leading to the next sequence.   (Here I'm happy to echo Richard Kugler.) The foreign policy "peaks" of the Twentieth Century I see as the First World War, Second World War, and the end of the Cold War.

Consider the massive reordering that we subsume under the shorthand of the First World War: Ten million war dead, innumerable other casualties, the birth of the Soviet state and the surge of international communism, the spread of European revolution, and the great influenza pandemic. The metaphor of the tinderbox is the time-honored one for that period. The new description that makes sense, however, is of a collection of factors building to criticality. There is a relatively minor event—the assassination of the Austrian Archduke by a Serb—leading to disproportionate outcomes and massive reordering. So too with the 1939-1945 period. The stage of building to criticality was increasingly evident from 1931 onwards, and we are all familiar with the spectacular change and the fundamental reordering that followed.

The collapse of the Soviet Empire is the third instance of global critical change that I'd highlight. The consensus I detect in the conference is that we really do not understand the post-collapse period. The East-West struggle kept a lid on conflict. Communism suppressed the destabilizing phenomena of nationalism and crime by making these, as so much else in the society, state monopolies. Ukrainian nationalism was illegal; Soviet patriotism was mandatory. Criminal gangs were strictly suppressed in the USSR; the *nomenklatura*, in contrast, was a powerful Cosa Nostra of its own. Now, with the Cold War ended, the state's monopoly on these enterprises has ended and we are dealing with the unpleasant sequelae of freedom, whether in Chechnya, the Balkans, Nagorno-Karabakh, or in the rise and spread of the Russian Mafia. In terms of our theory, the degrees of freedom in the system have greatly increased.

There's another way of viewing this, however: the fact that the great Cold War struggle diverted us from the accelerating chaos, the true dynamism, in the world and only now are we perceiving the scale of the global challenges: in terms of environmental disaster, water shortages, climate change, dysfunctional national cultures, and the breakdown of the nation-state. The response to this reordering is by no means complete, and this is a complex and intriguing area.

In each of these three cases, World War I, World War II, and the end of the Cold War, we were unprepared for the sequence and the magnitude of events. Nations at rest want to believe they will remain at rest.

By far the most interesting aspect of strategic criticality to me as a practitioner of foreign affairs is the policy
response to it, particularly the U.S. response. The
fundamental response to the chaos of these events
was a great premium on building stability: imposing
form on the chaos of nature. This is completely understandable. The first two critical reorderings of this
century were spectacularly painful. And certainly
people have a powerful drive toward stability. We thrive
within a clearly defined range of economic and political parameters.

The converse of this is that we perceive the chaotic
as at heart threatening. For proof of this, we need not
look at the upheavals of this century, but let's go back
to the fundamental level of dynamical systems theory,
the mathematical. Mandelbrot, in his wonderful book,
The Fractal Geometry of Nature, describes the Cantor dust and terms it "another awful mathematical
object ordinarily viewed as pathological." Further, he
notes that "many writers refer to [the graph of the
Cantor function] as the Devil's Staircase." We find this
same genre of mathematical objects referred to as "a
gallery of monsters"; Mandelbrot himself creates a
"fractal dragon." The irregular, the discontinuous, the
extraordinary is threatening.

So on the policy level, a consistent response of Western policy makers to discontinuous, chaotic,
threatening events has been to apply the arts of strategy. I am not indiscriminately criticizing the desire for
structure. But I think it's very important that we be
aware of this and watch for this powerful tendency in

ourselves, corporately, as policy makers. Thus we saw the great efforts by Western policy makers to devise a stable structure for international affairs to guard against the possibility of any recurrence of these events. After the catastrophe and reordering of World War One, we had The New Diplomacy, as it was termed: the ambitious attempt to found a League of Nations, the Washington Naval Conferences, the creation of a World Court, very broad Geneva disarmament talks, and of course, the Kellogg-Briand Pact for the Renunciation of War. There's an interesting sidelight here; this attempt to tame chaos on the international front was mirrored by the drive for stability—"the return to normalcy"—in domestic affairs. Prohibition is the outstanding example of this: activists attempted to tame the turbulent in domestic life as diplomatists attempted to tame the turbulent across the seas. Actually, I saw a contemporary magazine article pointing out the similarities and observing that from the European point of view, the Kellogg-Briand Pact was the same as the 18th Amendment, but without any Volstead Act to enforce it. Both Kellogg-Briand and Prohibition failed, of course. The paper strictures of well-meaning diplomats, up to and most of all, in Munich, were no match for turbulent reality.

After the catastrophe and reordering of World War Two, we saw American leadership in the formation of the United Nations and military pacts. In terms of international affairs, the decade of the fifties was far less turbulent that the decade that preceded it or for that matter, the one that followed it. Certainly there were many smaller upheavals—smaller avalanches, of a sort, in that period. Yet critical reordering—massive upheaval—in our century is followed by a period of relative quiet. This is not a profound point. What I want us to be aware of is how powerful the drive is for stability in the wake of the massive change—and the fact that the self-ordering of international affairs builds inevitably toward the next reordering. All stability is metastability, temporary stability.

Events in recent years give weight to the argument that the American response to chaos is the desire for structure. In homage to Professor Gell-Mann and his *CLAW* concept, let me call this part of my talk *SLAW*, or Sharpened Lunge Against Wishfulness. Let me explore this with a look at the end of the Soviet Union.

When did the catastrophic collapse of the Soviet Union begin? In November 1989, with the fall of the Berlin Wall? In June 1990, with the declaration of Sovereignty by the Russian Parliament? The rise of republic power in January 1991? Inarguably, however, the coup attempt of August 19, 1991 was a pivotal point. And at this critical point in history, the first responses of the White House in favor of structure are archetypal in disclosing the underlying drive for stability. At a news conference after the reports of the coup were received, President Bush confronted the turbulence by saying, "We expect that the Soviet Union will live up fully to its international obligations." And then, "There is very little that we can do now." And he referred to Gorbachev in the past tense, and made it clear that the U.S. sought the maximal degree of stability in the midst of this sharp change.

The discomfort with chaos and the drive for stability underlie those unworthy comments. Later the White House got it right, but I find it very revealing that those initial comments disclosed the attachment to immediate, stable order. If you look at the subsequent implementation of Soviet policy, it was cautious, timid, backward-focused. I can't help but contrast those comments with some earlier comments of Gorbachev when he was asked how he would like future generations to describe his contribution to his country. His answer? *Dinamichestvo*, dynamism.

So too with the end of the Gulf War. The premature termination of the Gulf War had many factors, but near the top were nods to the normative, the desire not to appear as a bellicose, "uncivilized" state by destroying the Republican Guard and the desire not to exceed the terms of the relevant Security Council resolutions. Those nods have proved costly in the succeeding years.

As I said earlier, we cannot appreciate the actual post-Cold War period yet, but we see indications of the shape things are taking. Let me point to the resurgence of the idea of international tribunals, in Bosnia and Africa. In Africa, the idea of such a tribunal represents the triumph of wishfulness over reality. I read in the *Post* this week of the lack of progress made by a War Crimes Tribunal on Rwanda. That this War

Crimes Tribunal is painstakingly sorting through the evidence and looking toward future judicial proceedings is less the stuff of realistic foreign relations than of satire, and I know I'm not being diplomatically correct in saying this.

I can't be completely hard on the structures of legality here, for two reasons. One is that in the long term, we do indeed need to develop an accepted code of behavior in both international and domestic spheres. That is a worthy goal, and one which is both the cause and the effect of the prosperity and stability that prevails in much of North America, East Asia, and Europe. A second reason that I can't be too hard is that we may have nothing else to offer in tragic situations but the palliative of legal measures, the fiction of international standards. The Central African situation is a case in point. Perhaps there is nothing effective that we are able or willing to offer to peoples who are locked in bitter, intractable conflicts.

But the basic point I want to get across here is that we have to be illusion-free when it comes to the limits and the appropriateness of international law and similar structures. The long-term goals of international law are worthy ones, but we always have to count the costs of what we pay for that in the short term. There is another facet to this desire for stability that concerns me. It is the spread of peacekeeping forces and the increasing use of the U.S. military as peacekeepers. I see two basic uses for peacekeeping, one productive and one that makes me wary. The first, the productive use, is the use of peacekeeping troops in implementation of a stable solution to a crisis, as in Cambodia. The second is the use of peacekeepers as placeholders: freezing the sides and clamping a lid on conflict. It's hard to argue against a pause in bloodshed. But if peacekeeping becomes an end in itself, as a means of achieving a pseudostability, and thus forestalls moving a crisis toward tough decisions needed to achieve a stable end, then we're on dangerous ground. The fact is that if we want to solve the problems in much of the developing world, we need to be looking at intensive, intrusive remaking of the societies. That may well be too much for any nation or group to take on. But that, I think, brings us closer to the true nature of the problems.

I also wonder about the effect on the military's view of itself that peacekeeping has. Here again, we pay a price. I have a traditional view of the military's mission: I believe our force exists to destroy the opposing forces. I am concerned that peacekeeping missions as a mission norm will over time blur and blunt the capabilities of our forces and give us too many people in the uniformed ranks concerned with political subtleties and nuances. We have too many of those in the State Department now, why do we want more at DOD? Remember Bob Jervis's discussion of unintended consequences.

So what do we do when faced with an environment of self-organized criticality? First, we need to understand the nature of the environment we face, go against our cultural patterning, and recognize that not all chaos is bad and not all stability is good. The Soviet coup is the case in point for this. We should also be aware of our tendency to create myths in service of our beliefs. Charles Krauthammer refers to the "Myth of multilateralism." This is one of them. Another is the belief in common international values. I am reminded of a cartoon from the Soviet days showing a haggard prisoner being dragged before a commissar. The caption read: "Sure we are believing in human rights, comrade. First you must prove you are human." We should be clear about the price we pay for stability and service to assumed international norms.

A sidelight here that will be no surprise to anyone who has worked in Washington is that policy making is chaotic. More precisely, it is turbulent and weakly chaotic, exhibiting features of self-organized criticality. George Shultz said that in Washington, no issue is ever truly settled. There's a deeper truth in that remark—policy decisions are metastable.

Unfortunately, description of the environment does not mean prediction. Description does not mean prescription either. The fact that we now see the world as subject to criticality doesn't really tell us how to use that fact. One prescription I have, though, is that we need to be open to ways to accelerate and exploit criticality if it serves our national interest, for example, by destroying the Iraqi military and the Saddam state. The key is national interest, not international stability. Indeed, we already push a number of policies that accelerate chaos, whether we realize that or not:

promoting democracy, pushing market reform, and spreading, through private sector means, mass communication.

Another prescription I have is to pay serious policy attention to environmental and resource issues. A few months back, I was on the Indo-Pakistani border, at Amritsar. The talk there was not of armed hostilities or Kashmir. It was on the plummeting of the water table. Now, what effect will that have in 15 years on Indo-Pakistani hostility? I have been much influenced by Robert Kaplan's writings on "The Coming Anarchy," and we need to reorient our policies to take account of these issues. Criticality tells us that all stability in a critical environment is metastability. One implication is that we cannot take the continued stability of the United States as invariable.

I have been critical of some of our reactions to chaos here, because at heart the international environment is conflict-based, and we ignore that fact at our peril. The world as international arena is a correct metaphor—and there is often no law in the arena.

I lifted that phrase from Camille Paglia, who lifted it from the movie *Ben-Hur*. In the art of foreign policy, this phrase makes more sense than we care to admit. Our reaction to criticality is a byproduct of our culture. Structure and culture are at war with the chthonian. It is noble for us as policy makers to try to triumph over chaotic nature and impose the art of diplomacy, of war, but a prerequisite for this is seeing the world as it is, not as we wish it to be.

Part Three

Complexity Theory, Strategy, and Operations

Clausewitz, Nonlinearity, and the Importance of Imagery

Alan D. Beyerchen

O ne reason for historians to play a role in national security affairs is that the narrative understanding of the past offers a reservoir of experience upon which to draw. This is, of course, common practice and common sense. In an era of significant transition such as the post-Cold War period in which we live, most people rely on their sense of the past to orient themselves and gain a feel for the direction of developments.

Many respected commentators argue that today we are on the cusp of the demise of the nation-state as the primary actor on the global political stage. The rival actors, according to Carl Builder and many others, function at both larger and smaller scales of

organization.[1] The European cliché says that authority is leaking from the nation-state at both the top and
bottom: the supra-national structure of the European
Union vies with sub-national levels of government and
cross-national regions for the allegiance and energies
of leaders and populations. On a global scale, national boundaries are overspread by multinational
corporations, transnational criminal organizations,
non-governmental organizations and religious authorities and sects. Meanwhile, ethnic groups, local
organizations and neighborhoods carve out increasingly defiant enclaves.

It seems to me, however, that entities at both larger
and smaller geographical scales will continue to have
need of the nation-state, even as the number of perceived "nations" and constituted "states" multiplies.
Some actors will want to retain it as a shield behind
which to conduct their activities. Others will depend
upon it as a base of operations or as a source of resources upon which they will prey. Still others will need
it as a convenient target of their rhetoric in order to
galvanize action among their followers. And some
states with no national constituency and some nations
with no state at their disposal will continue to avail
themselves of the symbols and practices of nation-
states for decades to come in order to legitimate their
claims to existence. Prior to the demise of the nation-
state those that exist are likely to fragment and multiply,
while maintaining the trappings of authority in an increasingly complex "inter-national" arena. During this
transition period, nearly as harrowing as the nuclear

proliferation we are facing is the national proliferation that will accompany it. Then will come the post-nation-state era.

Part of the historian's function is to explore the long-term view of the past in an effort to minimize temporal myopia. The nation-state is not likely to last forever—nothing really does, because entities either adapt to change and thus at some point become significantly different from their earlier incarnations, or they fail to adapt and disappear (with or without trace). But the nation-state is also not likely to evaporate in the next congressional budget cycle. After all, its demise or "withering away" has been projected by one observer or another from the mid-nineteenth century onward. It will still be around for a while.

Clausewitz as Theorist of the Nation-state and of War

The modern state has its roots in the secularizing tendencies of the late Renaissance and the onset of early modern warfare in the seventeenth century. The modern nation-state came to prominence with the French Revolution in the 1790's. Although not usually portrayed as such, an important theorist of this form of government was Carl von Clausewitz, who understood the energy unleashed in the emotional calls to arms of large portions of the male citizenry in Europe during the Revolutionary and Napoleonic eras.[2]

Clausewitz realized that the radical transformation of the scale and nature of war in his lifetime was due to a deeper phenomenon. This was the new participation of the citizenry as a whole in <u>politics</u>, a participation that characterized the transition from the modern state to the modern nation-state. Broadened political participation was at the heart of the French Revolution, Napoleon's successes, and also—ultimately—the measures adopted by Napoleon's opponents in order to defeat the French. Clausewitz understood political participation as stimulus for, exercise of, and constraint upon <u>power</u>. He knew that neither the Revolution nor the reforms created to combat it could be rolled back for long, because, as he wrote in his manuscript *On War*, "... once barriers—which in a sense consist only in man's ignorance of what is possible—are torn down, they are not so easily set up again."[3]

Thus the devolution of power—the democratic, egalitarian or fragmenting trends we have heard so much about at this conference—are related to the development of the nation-state itself and the continuation of broad trends that created the context for all Clausewitz's attention to the phenomena in and of war.

Clausewitz was also a theorist of war, which he perceived as a nonlinear phenomenon.[4] In order to discuss his views let us start where he does, as a good theorist, with definitions. In his work *On War*, Clausewitz first says that war is a "duel." This usually generates the image of two independent opponents crossing swords with one another or firing pistols at twenty paces. Actually this is too discrete and linear an understanding. The word which is usually translated as duel is *Zweikampf*, which literally means "two-struggle." The image Clausewitz himself offers on the same page is in contrast quite nonlinear: two wrestlers struggling with one another. The (presumably Greco-Roman and not WWF) wrestlers interact, generating positions and shapes that neither could possibly create alone.

Clausewitz also holds that war is the "continuation of policy" by other means. The conventional approach to this definition envisions a compartmentalization of politics (*Politik*, which also connotes policies) and war in a linear sequence—first comes politics/policies, then war, then politics/policies again to make and maintain peace. Furthermore, these interpretations hold that *Politik* drives war, but not vice versa. Actually the German word we translate as "continuation" (*Fortsetzung*) means literally a "setting-forth." This term does not require a sense of leaving something behind in the process; only our linear preconceptions lead us to imagine a norm in which the conduct of war is insulated from its context. A different approach emphasizes that Clausewitz believes war is not linear: war is a subset of the political context, and, furthermore, politics and military action interact in a complex, continual feedback process. As is the act of going to war in the first place, <u>every</u> act in war is the "setting forth" of politics/policies.

Furthermore, the conduct of any war affects its character. How else could Clausewitz have conceived the relationship between war and *Politik*, given his understanding of the new relationships created by the nation-state? New tactics and technologies affect the way a war is fought. But consider also the ways in which the Prussian state was forced to undertake deep political and social reform in order to respond to the changed demands of the battlefields of the time, and the ways in which those reforms affected the structure and combat characteristics of the Prussian armies in the field. Experience told Clausewitz that the conduct of war affects its political context, which responds with changed parameters and goals that alter the conduct of war, which affects the political context anew, and so forth.

Finally, Clausewitz claims that war is a "remarkable trinity" composed of the primordial passions of the people, the rational policies of the state, and the combination of incidents in battle (good luck or bad luck, the genius of the army commander, accidents with great consequence, etc.). Theory, he says, should be treated as if it were an object suspended among these three points of attraction. Many commentators have taken Clausewitz to mean that war should be treated in linear fashion in the form of a triangle, with lines bisecting each angle to create a static intersection point at which theory resides. But actually, the word translated as "suspended" (*schwebend*) connotes a hovering or a floating about. The physics demonstration of a pendulum tracing out a highly complex and irreproducible trajectory among three magnets is exactly what Clausewitz had in mind. And it is the quintessential demonstration of a nonlinear system highly sensitive to the initial conditions under which it operates.

Every war involves inherent nonlinearities that pose problems for prediction, and Clausewitz talks about three broad categories of nonlinear factors that make for unpredictability in war. The first is interaction between animate entities that act, react and even preempt. This is not a simple binary opposition, for to Clausewitz much of what matters takes place in the spaces between and around the interacting entities (hence the image of the wrestlers or magnetic fields). His attention is always drawn to where boundaries are complex rather than simple.

The second source of unpredictability is what Clausewitz chooses to label "friction." We must keep in mind that this was a term taken from the research forefront of his own day, a high-tech notion from the emerging science of thermodynamics. Clausewitz had in mind that wars are dissipative systems, which in the real world (as opposed to that of pure theory) always suck in and consume people and other limited resources. In another sense he meant with this term the amplification of a microcause to a macro-consequence, in a kind of cascade of things gone wrong. This is his more interesting version of the adage that "for want of a nail the shoe was lost, for want of a shoe, ..."[5]

Clausewitz also regards chance as one of the sources of unpredictability in war. He nowhere offers a concise definition of chance, but it seems to me that he addresses three forms of chance in *On War*. The first is stochastic phenomena, because Clausewitz repeatedly emphasizes that there are no firm boundaries that isolate war from its political context. Another is the amplification of undetectable microcauses, which ties chance and friction together in the inevitable confusion of war. And a third is the set of analytical blinders we unavoidably wear in real life, blinders that make us slice up the universe in manageable pieces and then perceive as chance the intersections of some of those slices.

None of this means that linearity cannot ever be achieved in war, but it does indicate that linear, predictable relationships are hard to come by. They are also always attained at some significant cost. More importantly, our search for and reliance upon proportional and additive relationships creates a set of those analytical blinders that constitutes a potential weakness available to our opponents. The purpose of any theory of war for Clausewitz is to explore the entire range of possibilities, including counterfactuals in the sense that physicists understand them. It is not to generate a preconceived set of stable relationships, a checklist of laws valid upon any occasion, "since no prescriptive formulation universal enough to deserve the name of law can be applied to the constant change and diversity of the phenomena of war."[6]   Instead, theory should be guided by knowledge of past human experience and the best current scientific understanding of reality and natural constraints. According to Clausewitz, history must inform theory and serve to educate the commander. Only in this way can the nonlinear nature of war be understood adequately. This is the import of the images Clausewitz uses so astutely.

About Nonlinearity and Imagery

Why harp on nonlinearity, much less imagery? Why do they matter? Let us start with nonlinearity.

One reason for emphasizing nonlinearity is that it constitutes the well-established mathematical property underlying and making coherent all the faddish-sounding new sciences: deterministic chaos, fractals, self-organizing systems far from thermodynamic equilibrium, complexity and complex adaptive systems, self-organizing criticality, cellular automata, solitons, and so forth. It was in various ways sensed by the ancient Greeks. Newton understood it, although the great French mathematicians of the eighteenth century linearized Newton as they popularized his ideas—much of what we decry as "Newtonian thinking" would actually be better ascribed to Laplace. Clausewitz recognized its importance as an alternative to Laplacian precepts, perhaps because he had such great antipathy toward those things that were French. Yet no one before the late twentieth century could solve the interesting problems posed by many nonlinear equations. There are no analytical techniques that work well, and numerical methods were just too cumbersome and time-consuming. Most scientists just bracketed out the nonlinear elements of their equations and went with the idealized linear approximation. Now computers allow us to go after formerly intractable problems by pursuing numerical solutions.[7]

The connotations of linearity still drive a great deal of our thinking, especially in mechanics and the many social scientific disciplines that implicitly try to copy the success of mechanics. Linearity offers structural stability and emphasis on equilibrium. It legitimates simple extrapolations of known developments, scaling and compartmentalization. It promises prediction and thus control—very powerful attractions indeed. But linear systems are often restrictive, narrow and brittle. They are seldom very adaptive under significant changes in their environment (as Clausewitz clearly understood). Bureaucracy is the quintessential linearization technique in social affairs.

The connotations of nonlinearity comprise a mix of threat and opportunity. Nonlinearity can generate instabilities, discontinuities, synergisms and unpredictability. But it also places a premium on flexibility, adaptability, dynamic change, innovation, and responsiveness. This is why there seems to be serious metaphorical value in the images and ideas emanating from the new sciences.

Murray Gell-Mann, James Rosenau, and others caution wisely against expecting too much, too soon from the new sciences and stress the informed use of metaphor for now. I could not agree more. But if this sentiment implies that metaphors are merely poor substitutes for adequate models, then I could not disagree more. Metaphors are extremely powerful in their own right and should not be treated simply as tokens along a tollway toward models.

What is metaphor? Is it only a stylistic flourish, as most of us think who encountered metaphors primarily in literature classes in school? No, metaphor is much more significant, as philosophers and linguists are beginning to demonstrate more and more convincingly.

A metaphor is usually a statement that is paradoxical. It is literally false according to the rules of abstract rationality (i.e., logic, truth tables), but is true according to the rules of imaginative rationality (i.e., art). Metaphor constitutes a ubiquitous, irreducibly complex aspect of any natural language. It is an essential "AS" gate in our cognitive processing. It is a crucial way we understand one thing as another.

Metaphors are embedded throughout our speech patterns (including the word "embedded" here). They are jarring when new, but often we use "dead" metaphors or clichés such as the wings of a building, the branches of science, weighing our options, or sitting at the foot of a mountain. Each such "gate" is much more than a word. Contemporary researchers tell us that metaphors are indicators of networks of meanings and entailments that dilate or constrain both our perceptions and our conceptions.[8] It is furthermore possible to extend this understanding to visual and other metaphors such as the Mandelbrot set that enlivens our program cover at this conference.

The importance of metaphor has long been understood. Aristotle wrote, "The greatest thing, by far, is to be a master of metaphor. It is the one thing that cannot be learned; and it is also a sign of genius." He contended that it is so indicative of power that is it not appropriate for slaves to use it. Hobbes took a related but different tack. For him, metaphors were dangerous not due to their power, but their tendency to confuse us as "senseless and ambiguous words." He distrusted reasoning with metaphors as "wandering amongst innumerable absurdities." But this was the same Hobbes who also wrote: "Why may we not say that all <u>Automata</u>...have and artificial life? For what is the <u>Heart</u>, but a Spring; and the <u>Nerves</u>, but so many Strings; and the <u>Joynts</u>, but so many <u>Wheels</u>..."[9]

This is quite arresting and interesting. It could be mere sloppiness on the part of Hobbes, but in the writing of so powerful a thinker something else may be at work. That something is also displayed in the words of Clausewitz. Critical studies, he says, are imperiled by narrow systems used as formal bodies of law and "a far more serious menace," the "retinue of <u>jargon</u>, <u>technicalities</u> and <u>metaphors</u> that attends these systems. They swarm everywhere—a lawless rabble of camp followers."[10]

To condemn metaphors in such a colorfully metaphorical way implies that Clausewitz thought—as did Hobbes—in profoundly metaphorical terms. Think merely of his "friction," or "fog" of war, or "center of gravity." Recall how a defeat "leaves a vacuum that is filled by a corrosively expanding fear which completes the paralysis. It is as if the electric charge of the main battle had sparked a shock to the whole nervous system of one of the contestants." Or how routine constitutes a clock "pendulum" that reduces natural friction and "regulates" the mechanism of war. Or how war has its "own grammar," but not its own logic. Or that politics is "the womb in which war develops— where its outlines already exist in their hidden rudimentary form, like the characteristics of living creatures in their embryos."[11]

Why did Clausewitz resort to this "lawless rabble of camp followers" in his own language? One reason was that he wanted to draw upon history to generate theory. In historical studies a major goal is frequently to understand one thing (the present or a vision of the future) in terms of another (the past). Metaphor is very robust for this purpose. Consider the staying power of the metaphor of the 1938 Munich agreement in American foreign policy since World War II. To claim some action is necessary to avoid "a Munich" is to offer a justification of enormous magnitude; to claim some other course will lead to "a Munich" is to denounce its proponents in the most damning terms as appeasers. Metaphors appeal to imaginative rationality and often evoke indelible images.

Clausewitz also wanted to draw upon theory to better understand history and the power of our narratives of the past. We need only think of the efforts of his contemporary, Hegel, to recognize this desire as part and parcel of the age. History was viewed as conceptually akin to the biological and geological sciences of the age. It was an exercise in taxonomy that would soon lead to a new and bolder understanding of ourselves and the world we inhabit.

Yet another reason Clausewitz relied upon metaphorical imagery was that he did not trust the established
jargon of his day, which was full of rigid (and French!)
geometric principles and models. He preferred the
new sciences of his time—chemistry, thermodynamics, magnetism, electricity, embryology. These offered
novel, high-tech, research-forefront terms for the dynamic phenomena he wanted to discuss. Analytical
models can be superior devices in efforts to understand the logical consequences of our assumptions.
Their appeal resides largely in their beauty and utility
as a form of controlled experiment, especially for
modeling phenomena that can be controlled in turn.
Yet these models, too, draw upon linguistic structures
that we too often associate with literature alone—the
tropes of metonymy (allowing the attribute to stand
for the whole) and synecdoche (allowing the part to
stand for the whole). The attributes we tend to call
variables, while the model itself is a scaled-down version of the system we want to investigate. Everything
hinges on the assumptions we build into the model.

Clausewitz appears to have understood that metaphors can be superior when the phenomena of interest
cannot be controlled, or you are unsure of the necessary assumptions. As evolving things, metaphors are
open to novelty, surprise, inspiration and even mutation. They therefore can capture the underlying
processes of other evolving entities surprisingly well.
If the metaphors are really successful, of course, they
may become mere commonplace, frozen images that
get passed along unthinkingly and thus constrain our

imaginations. But this is also part of the way evolution works. Metaphoring (as opposed to traditional analytical modeling) is a process of exploring some interesting possibility space with contingency and feedback. Each biological mutation is such an exploration, as is each historical event. This is a crucial aspect of Clausewitz's method of analysis and his approach to war.

Conclusion

What is the utility of thinking about war—for our potential opponents and ourselves—in nonlinear terms, especially in the high-tech, research-forefront metaphorical terms from the new sciences? For our opponents the usefulness may be the same as it was for Clausewitz. The Germans were underdogs to the French, and Clausewitz wanted to understand and use against the French their linearizing blindspots. He also needed to be the champion of disproportionate effects and unpredictability, for in a linear, predictable world Prussian resistance to Napoleon after 1807 was futile. The opponents of the United States will be looking for our blindspots in an effort to seize opportunities to surprise and shock us. They may also be able to compensate for their disadvantage in military confrontations such as the Gulf War by consciously striving to affect the political context in order to change the conduct of warfare. An understanding of the porousness of the boundaries between politics and war can be a real weapon against those who envision those boundaries to be impermeable.

We need for our own sake to understand the limitations our imagination places upon us. Linearity is excellent for the systems we design to behave predictably, but offers a narrow window on most natural and social systems. That narrowness sets blinders on our perception of reality and offers a weakness for an opponent to exploit. But if we know our limits, we can minimize the extent and duration of our surprise, reducing its value to someone else. And an expanded sense of the complexity of reality can help us be more successfully adaptive amid changing circumstances. By thinking more constructively about nonlinearity, we might be able to design more robust systems when we need them. A new form of modeling that takes such concepts as self-organization to heart allows structures to bubble up from below rather than be imposed from above. With such tools we might come to understand better the biological and historical processes with which we must deal. And we may come to realize how conventional, analytical predictive techniques can themselves stimulate a self-defeating, unfulfillable desire to control more of the real world around us than is truly possible.

In his opening address at this conference, Murray Gell-Mann was right. The issue is not that we lack information about the world; it is that we need better schemata. We do not know enough about the new sciences to apply them very well yet, but every attempt helps us learn and adapt to the changes with which we must cope.

End Notes

1. See, for example, Carl H. Builder, "Is It a Transition or a Revolution?" *Futures* (March 1993); Samuel P. Huntington, *The Clash of Civilizations: The Remaking of World Order* (NY: Simon & Schuster, 1996); Alvin and Heid: Toffler, *War and Anti-war: Survival at the Dawn of the 21st Century* (Boston: Little, Brown, 1993).

2. On this point see especially Peter Paret, *Clausewitz and the State: The Man, His Theories and His Times* (Princeton: Princeton University Press, 1985; originally published 1976).

3. *On War*, edited and translated by Michael Howard and Peter Paret (Princeton: Princeton University Press, 1976), p. 593.

4. This section is based on my article "Clausewitz, Nonlinearity and the Unpredictability of War," *International Security* 17 (Winter 1992/93): 59-90.

5. Along these lines, a very intriguing exploration of what Clausewitz meant by friction and what the term means today has been recently offered by Barry Watts in his *Clausewitzian Friction and the Future of War*, McNair Paper 52 (Washington: Institute for National Strategic Studies/ NDU Press, 1996).

6. *On War*, p. 152.

7. A sense of the contrast between the two techniques is offered by Larry Smarr, "An Approach to Complexity: Numerical Computations," *Science* 228 (26 April 1985): 403-08.

8. For a very readable exposition, see George Lakoff and Mark Johnson, *Metaphors We Live By* (Chicago: University of Chicago Press, 1980). For a variety of current approaches, see Andrew Ortony (ed.), *Metaphor and Thought*, second ed. (Cambridge: Cambridge University Press, 1993).

9. For these and other passages about metaphor, see Gemma Corradi Fiumara, *The Metaphoric Process: Connections between Language and Life* (London: Routledge, 1995), here pp. 1-5.

10. *On War*, p. 168.

11. *On War*, pp. 225, 296, 605, 149 respectively.

Complexity and Organization Management

Robert R. Maxfield

In recent years there has emerged a collection of interdisciplinary scientific efforts known as the science of complex systems, stimulated by the pioneering efforts of the Santa Fe Insitute. Complex systems, which consist of many interacting entities and exhibit properties such as self-organization, evolution, and constant novelty, exist in all the domains of our world—physical systems, biological systems, human social systems—and are very difficult to comprehend by the standard reductionist analytic approach of modern science. The science of complex systems attempts to discover general laws governing such systems by bringing together people and ideas from many disciplines. As yet such general laws have not been found —indeed completely general laws may not exist—but the efforts have yielded much deeper insights into the systems studied. The scientifically significant results are so far mostly in the physical and biological domain, but the metaphors have proven to have tremendous appeal and utility in studying humans and human social systems.

The basis of the appeal of complex systems metaphors in thinking about our human world is not hard to find. We live in a time of rapid, unpredictable, and novel change; the manner of the demise of communism is an example that captures the essence of unpredictable change. For those of us with responsibility for effectively managing organizations, whether in the private or public sector, the instabilities in our present world call into question most of the conventional wisdoms about management.

My purpose in this paper is to propose that complex system metaphors provide a valuable intellectual framework for thinking about our human world and managing the organizations which comprise it. My perspective is as a practitioner of management in the high-tech industry, arguably the industry that has undergone the most rapid and fundamental changes over the last 40 years. I will try to impart some of the insights I have gained by applying the framework of complex system metaphors to my experience over more than 25 years. Although insights gained from the high-tech part of the private sector may appear at first glance appear to have no applicability to other domains such as the military or foreign policy, I believe at the proper level of abstraction all human organizations and institutions have much in common.

Since the study of complex systems is a recent development, most of us were trained in other fields, and when solving problems we apply the "arbitrary" component that Thomas Kuhn [1] refers to in his seminal work on scientific revolutions. For example, those of us who approach the world from a systems engineering perspective bring to it a background rich in mathematics, system theory, linear systems analysis and control theory, as well as a knowledge of decision analysis and game theory. Needless to say, with this kind of background, one tends to look at problems in a "systematic" way, trying to identify relevant and controllable variables, to decompose the problem into manageable parts, and to formulate the problem in terms of the solutions tools and approaches that are our stock in trade.

Sooner or later we come across a problem or set of problems that is not tractable by applying the "standard" approaches and tools that came with our selected profession. For me this happened sooner rather than later. In 1969, shortly after I completed my engineering doctorate, in which I emphasized systems theory, I succumbed to Silicon Valley fever (though the term "Silicon Valley" had not yet been coined) and co-founded a computer company, ROLM Corporation, with three other colleagues, all with similar backgrounds which included almost no management experience or business education.

Eager to bring the tools of my profession to bear, I initially tended to apply my systems training to managing an organization. Need to make a decision? Apply decision analysis; define all the possible consequences of all the possible actions one might take, then assign probabilities and value functions to these various outcomes, then compute expected values and ascertain the "optimal" decision. Worried about competition? Apply game theory. It did not take long to realize that this "engineering" approach to problem solving was unsuited for the rapidly changing environment which I was in. Fortunately, my partners had sufficiently different perspectives and skills that as a group we were able, with plenty of trial and error, to manage and grow a human organization operating in a rapidly changing external environment.

Over the next twenty years, the company successfully grew to over 10,000 employees, but I never really felt comfortable with many aspects of organizational management. I acquired a set of skills, tools, and techniques that tended to work, but I had no overall intellectual framework, or mental model, for thinking about the world in which I was embedded. Several years ago, pursuing an interest in economics, I became aware of the Santa Fe Institute through one of its first publications, *The Economy as an Evolving Complex System* [2], and was introduced to the emerging field of the scientific study of complex adaptive systems. I became convinced that a complex systems approach could provide the unifying intellectual framework for thinking about the world of high-tech management. Just as a complex systems approach could show why current economic theories of equilibrium, perfect rationality, and decreasing returns are incapable of understanding or explaining the 20th century global economy, it could also show why decision analysis and game theory are inadequate to explain or prescribe the behavior of firms.

In this paper, my objective is show how complex system metaphors can be used to gain perspective on the world of high technology, and to suggest some implications for management of any organization operating in a context of rapid change. I will first review the four major properties of complex adaptive systems and relate them to the high-tech world, then suggest by a "Darwinian selection" argument that there are some common attributes of successful organizations. Finally I will discuss strategic planning in complex environments.

Properties of Complex Adaptive Systems

By a *complex adaptive system*, or CAS, I mean an open-ended system of many heterogeneous agents who interact non-linearly over time with each other and their environment and who are capable of adapting their behavior based on experience. Open-ended means there is essentially limitless possibility for variability in agent characteristics and behavior. In non-human biological CASs, the source of agent variability is primarily genetic with inheritance; in human CASs the primary source of variability in behavior is the immeasurably large cognitive ability of the human brain. There are four major properties of the aggregate dynamics of CAS that set them apart from other systems: self-organization, evolutionary trajectories, co-evolution, and punctuated equilibrium. All of these properties are emergent, in the sense that complete knowledge of the individual agents is not sufficient to infer the details or timing of the aggregate properties. Professor Rosenau, elsewhere in this volume [3], has eloquently described these properties; I will briefly recap them and use them as a lens through which to view the world of high-technology.

Self-organization is the emergence of new entities or stable aggregate patterns of organization and behavior arising from the interactions of agents. Each higher level of organization has its own time-scale, and each new level has new kinds relationships and properties. That is, a complex adaptive system on one level is made up of lower level complex adaptive systems interacting and creating the higher level order. In human systems we usually take the lowest organizational level as the individual, although each individual could be considered to be comprised of lower level CAS, such as our brains and immune systems. Human CAS have several characteristics which distinguish them from other classes of CAS such as physical or biological systems. First, we have more levels of organization. The next level up from the human individual is family, clan, firm, etc. Going on up, we have on the economic side industries, regional economies, the global economy; on the governing side we have cities, states, nations. So there are multiple levels of nested complex adaptive systems in which humans operate individually and collectively. Second, every individual is usually a member of several higher level entities—family, employer, profession, church, city, etc. So self-organization is not strictly nested; complex webs of interconnections between human CAS exist at all levels. Third, the higher level (other than family) human organizations are social constructions as opposed to natural constructions. That is, the entity types are creations of our collective imagination to which we attach names, such as firm, industry, and economy. And the rules that determine the interactions between these entities are also socially constructed and are not fixed laws of nature.

Evolutionary trajectories means the future history of a given system from a given point in time can not be determined by complete knowledge of the present state, and if you "re-run the tape" many times, every trajectory will most likely be unique. In particular, "historical accidents"—he occurrence of certain *a priori* very low probability events—can dramatically change the outcome (e.g., Hitler's accession to power). However, in human systems as in simpler biological systems, the prerequisites for Darwinian natural selection are met—mechanisms for the creation of novel entities, limits to population of entities, differential entity survival based on relative fitness, and heritability of attributes—which ensures that in any given trajectory, we expect to see emergence of order in human systems analogous to the emergence of species and ecologies in nature.

Co-evolution takes the basic concept of Darwinian evolution to the next level. Instead of having a stable environment to determine fitness as agents adapt and evolve, a large part of each agent's perceived environment consists of interactions with other agents, who are themselves adapting and evolving. And each agent interacts not only with other agents at the same level in the organizational hierarchy, as when firms compete in an industry, but also with agents at higher and lower hierarchical levels, such as firms' relations with employees or the tax policies of the government. I believe that in thinking about human CASs, it is highly useful to include our *artifacts*—the inanimate things we create and make—as well as our organizations. In the term artifacts I include not only tools and products but information and knowledge. Our artifacts exhibit, in a more limited way, the properties of complex adaptive systems, in that they evolve (from the abacus to the personal computer), they co-evolve (weapon systems), and they exhibit increasing levels of organization (LANs to the Internet). And because our human organizations are largely organized around making and using artifacts, we really should view our human agents as co-evolving with the artifacts we create. The behavior of a particular agent depends, to a large degree, on the artifacts at its disposal. If, for example, a country has created a new weapon, its army will evolve to take advantage of the unique capabilities this new weapon offers. Further, if you are facing an army that has a different set of weapons, both the weapons you have and those that they have certainly matter, in terms of how you expect them to behave and how you are going to behave. Recently, the combination of two types of artifacts, weapons and computers, into a new type, smart weapons, has had an enormous impact upon defense systems at many levels.

Punctuated equilibrium is the tendency of a CAS to have stable patterns of activity for long periods of time, then have a short transition period of very rapid change in patterns, followed by new stable patterns of activity. In open-ended complex adaptive systems, it is usually impossible to predict when transitions will occur or what the resulting stable patterns will be. In our multi-level global human CAS, call it the *human world*, this phenomena occurs at all levels, and the question of stability versus instability depends on which part of the system you are looking at, what kind of patterns you are looking for, and what time scale you are using. For example, macro-economists studying the U.S. economy would say that since the 1940s the U.S. GNP has grown fairly smoothly over time, with a few blips here and there, and conclude the U.S. economy is in an equilibrium state, and liken it to a finely tuned, smooth-running engine of production. But if one drops down to the level of the firm, one sees thousands of firms going out of business every year, and new ones forming all the time, hardly an equilibrium state.

The High-tech Sector

If we take a centuries-long view of our human world, it is easy to see patterns of punctuated equilibrium. In the words of Peter Drucker,

> "every few hundred years in Western history there occurs a sharp transformation [in which] society rearranges itself—its world view; its basic values; its social and political structure; its arts; its key institutions. Fifty years later, there is a new world. And the people born then cannot even imagine the world in which their grandparents lived and into which their own parents were born." [4]

Most of the major transition periods coincide with the emergence of new classes of artifacts around which we reorganize ourselves—Gutenberg's printing press in 1455 driving the Renaissance, Watt's perfected steam engine in 1776 initiating the Industrial Revolution.

Unquestionably, the development of the digital computer in the 1940s, followed by the invention of the transistor about 1950 enabling the economic implementation of the computer, has spurred a new major transition phase for humanity, which many call the Digital Revolution. Over the past four decades we have seen many generations of evolution of new classes of artifacts enabling blindingly fast computation, unlimited information storage, instantaneous communication over vast distances. These capabilities are driving rapid changes in all aspects of our human world, but nowhere is the pace of change more rapid in the newly created sector of the economy— call it the *high-tech sector*—comprising those firms directly involved in the creation of the artifacts them- selves—computer and software companies, telecommunication companies, semiconductor manu- facturers, etc.

If we take as our frame of reference the complex adap- tive system consisting in the agents and artifacts of the high-tech sector, we can see an incredibly rich display of all of the properties of complex adaptive systems played out over 40-plus years. As an example of self-organization, we see first the emergence of the computer industry, followed by the software industry, followed by the data communication industry, each with its own identity, trade associations, trade shows, mar- ket research firms. Darwinian evolution is evident in the birth of new firms, typically inheriting "genes" of practices and cultures from older firms from which the entrepreneurs spun out, with survival of the fittest. Co- evolution is evident in the competition between firms leading to specialization into protectable niches, and in the entwined history of processor architectures, op- erating systems, programming languages, and networks. A good example of punctuated equilibrium is the "computer industry." First there was the era of the mainframe computer—room sized, costing millions of dollars. After an initial shakeout period in the '50s, there emerged stable market shares split among eight companies, with IBM holding over 70% of the total market. In the late '60s a new variant of artifact ap- peared, the minicomputer, costing tens of thousands

of dollars. This initiated a dramatic increase in the total computer market, and the minicomputer segment
became a very sizable fraction. A plethora of new companies, in addition to the existing mainframe
companies, vied for market share, but within a few
years the minicomputer segment stabilized, dominated
by four companies: IBM and three newcomers. Then
in the late '70s, yet another "species" emerged, the
personal computer, costing a few thousand dollars.
Again a spate of new companies emerged to compete in a vastly expanded market, in addition to existing
ones, and after a few years stability again set in with a
handful of companies dominating the market, all new
except IBM.

The *Economist* notes "twenty-five years ago only about
50,000 computers existed in the whole world; [today
there are] an estimated 140 million...and that does
not count the embedded processors inside cars, wash-
ing machines or even talking greetings cards. A typical
car today has more computer processing power than
the first lunar landing-craft had in 1969." [5] No matter
what metric you choose—mips per processor chip,
bits per memory chip, cost per mip, cost per byte of
memory, transistors per chip—performance has increased by a factor of 100 every 10 years for the past
three decades (Moore's Law). There is no reason to
believe that, at least for the next two decades, these
trends will change.

Characteristics of Successful High-tech Organizations

Clearly, organizations that survive and prosper in the high-tech sector must deal successfully with rapid change, not only in the artifacts with which they are associated, but in the agents with whom they compete and interact. Of the many thousands of new and existing firms that have attempted to compete in the high-tech sector over the last four decades, relatively few have succeeded. If we are searching for insights into managing organizations in rapidly changing environments, it would seem reasonable to look at these successful firms to see if they have traits in common - attitudes, management processes, organization forms, etc. Darwin's principle of natural selection would imply those traits or characteristics that confer the best fitness will tend to spread through the population, either by "inheritance" through spin-outs, or imitation by others of successful role models.[1] Are there such common traits? I believe there are, and I will try to summarize them here.

There are two key principles that high-tech organizations understand at a visceral level. The first is to recognize that *time is the scarce commodity*. An organization has to be able to match the rate of change in its environment. If it cannot, it does not matter what resources the organization has in terms of money,

---

[1] Admittedly we are dealing here with a relatively few generations compared to biological evolution, so my argument should be considered as suggestive rather than scientifically valid.

people, intellectual capital, goodwill, or any other re-source. An organization that cannot keep pace will inevitably fall farther and farther behind; having large resources will only prolong the death spiral. One metric crucial to many companies is the length of the product development cycle - the time between successive generations or major versions of a product. Thirty years ago, five years was an acceptable cycle. Twenty years ago an upper bound was three years. Ten years ago the best companies were shooting for less than two years. Now, a new buzzword in Silicon Valley is "Internet time" [6], with product cycles measured in months.

The second key principle is to recognize that *people are the key asset of any organization.* Why? Because people are the adaptive element of organizations. Learning and innovation come only from human cognition. Perhaps someday computers will exhibit true artificial intelligence, but that is a long time away, if ever. Humans are great at pattern recognition, great in making sense of "messy" situations, great at learning and adapting. The critical management task is to enable employees to most effectively use these capabilities to learn and adapt for the benefit of the corporation. High-tech companies have always been the leaders in attitudes, cultures, and policies to keep their employees motivated, happy, and productive. Few successful high-tech companies are unionized; a successful union organizing effort would be considered a catastrophic management failure. Unions create

adversarial relationships among classes of employees that deteriorate the potential for collective learning
and adaptation. If a significant number of employees
feel they are not getting a fair shake and need a union
to "fight" for them, management has failed.

If an organization takes to heart the two principles
concerning time and people, what else needs to be
done to ensure that an organization can adapt in a
dynamic, complex environment? I find it useful to break
this down into two questions: how can an organization *allow* adaptation, and then how can an
organization *encourage* adaptation. Let's consider
each of these.

Although humans are the adaptive element in every
organization, it does not follow that any organization
will be adaptive. In fact, there is a deeply embedded
metaphor in our society that works strongly against
adaptable organizations—the metaphor of the organization as a machine. The metaphor grew naturally
out of the last great social paradigm shift, the Industrial Revolution, in which science based on Newtonian
physics led to the development of machines that replaced humans and animals as sources of energy for
creating and transforming artifacts. A machine is a
system of carefully designed parts interconnected in
a precise way to accomplish a function repeatedly and
reliably. The key to a machine is that each part has a
known, predictable behavior in the system, and that
the interconnection of the parts results in the result
for which the system is designed. If one makes an
analogy to human organizations, in which human

beings are the component parts, there is the immediate problem that human behavior can be quite unpredictable. The answer to this, inspired by the work of Frederick Taylor [4] early this century, is to analytically determine the one best way to do each task, then train people to do it this way, and insist on reliable conformity—standard operating procedures. In a similar fashion, the interaction of the human components of the organization is carefully defined—who communicates with whom about what, who has responsibility for what. Since variability in results is to be avoided, authority to permit deviations from standard procedures is invested in only a few key individuals. We are all familiar with the end result of applying the machine metaphor—organizations that have precisely defined organization charts with many hierarchical levels, volumes of procedures defining most activities of the organization, and most major decision-making vested in a few central individuals at the end of long chains of authority. Staff organizations, mostly isolated from direct contact with the external environment, spend endless hours (aided by the writings of business school organization theorists) worrying about the "best" way to organize people into functional blocks, how these blocks should relate and communicate, designing "optimal" work flows and methodologies (aided by systems and operations research theorists). By their very design, such organizations do not allow for rapid adaptivity and innovation in response to external change. What capabilities they do have for change are vested in a very few people, rather than harnessing the cognitive capabilities of every member of the organization.

Suppose that, rather than using the machine analogy, we use instead the complex adaptive systems metaphor in thinking about organization structure and design, and view our organization as one CAS made up of many other CASs, namely the human members, and attempting to survive in an environment of many other CASs, with whom we must both cooperate and compete. Then, by the properties of such systems, we know there will be an inherent tendency for self-organization among employees, that continual evolution (read change) will be required in all aspects of our activities, that our external environment is not static but co-evolving with us, and that we can expect periods of very rapid change interspersed with periods of slower change. How then should we design our organization? Pretty clearly, it should be the antithesis of the machine-derived model. It should feature few rigid operating procedures, it should have great flexibility in organization structure, it should have widely delegated decision authority with short authority chains, and it should be very sensitive to changes in its external environment. These are indeed the features of successful high-tech organizations; in fact, I submit that these features now characterize almost all high-tech organizations as a result of Darwinian selection over many generations of evolution.

Suppose we were to study the organizational structures of two large companies, the first being an old-line type such as General Motors 30 years ago (before the Japanese ate their lunch), and the second a large high-tech company such as Intel or Microsoft. Both would have organization charts we could study, and on the surface they would appear similar, a hierarchical tree of sub-organizational blocks, although the high-tech chart would probably be much flatter. There would likely be an attached commentary describing the basic activities and responsibilities of each component sub-organization, together with an overview of how the components relate to each other. If we went to the managers of the components of the top-level chart and asked how their part of the organization is organized, they would produce similar structures. At this superficial level, we might conclude there are no real differences in the organizational structure of the two companies. But if we dug to a deeper level of understanding, we would find profound differences. If we asked to see the company procedures manual, the old-line company would likely produce a multi-volume set, and advise us that each component organization would have their own additional volumes. In the high-tech company we would be given a very slim volume that contained very few procedures ("you will do it exactly this way"), but instead mostly policies ("here are some overall constraints on the actions you can take") and guidelines ("here are some suggested ways to do it which usually work, but you are free to find a better way"). There would be a discussion of the company's mission and a discussion of the values that are expected to guide the behavior of all employees.

High-tech companies would consider it counter-productive to have highly detailed procedures for action and interaction; rather, they recognize that the formal organizational structure is just a *guide* for the kinds of relationships and interactions that need to develop for success, and that it is crucial to allow employees the maximum possible latitude for action. If we examined in depth the range of decisions managers at each hierarchical level could take without prior approval from a higher level or from peer levels, we would find it quite restricted for the old-line company but quite broad for the high-tech company, so decision making is broadly decentralized.

Rather than relying on a detailed formal organization structure to channel all activities and interactions, high-tech companies rely instead on the *informal organization*, the self-organizing networks of relationships that arise naturally from purposeful collective activity, and on *temporary organizations,* such as teams and task forces, for fast response to change. The informal organization contains collective wisdom about who has what skills and how best to solve problems. Further, it is fluid and adaptable. As conditions change, the informal organization rapidly deletes, modifies, and adds to the patterns of interactions in order to rapidly adjust to the situation. When a situation arises which strains the abilities of both the formal and informal organizations, rather than obsess about how to optimize the formal organization chart to deal with it, the best resources for dealing with it are marshaled from throughout the company, usually selected

via the informal organization, and a temporary organization is created and endowed with appropriate
authority, to determine and execute the appropriate
response. Sometimes, after the organization has responded to some challenge through temporary
organizational action, there emerges a realization that
a modification to the formal organization chart is appropriate for the changed context, but note that this
happens *after* the learning has occurred, not before.

Temporary organizations are not necessarily just *ad
hoc.* Very often they are routinely used for recurring
activities such as teams for product development
projects. Each time a new project is started, a team is
named with representatives from each relevant formal organization component, and the team is vested
with full responsibility for success of the project, then
dissolved when the project is completed. In most high-
tech organizations the concept of a team—small
groups of experts in their own domain, formed to work
together on a problem that requires expertise from all
their domains—is a standard organizational management tool.

Successful high-tech organizations view organizational structure and design as tools to help
organizations function, not as ends in themselves. In
rapidly changing environments, an organization should
have a toolbag of possible organizational structures
that can be called into play depending on the context.
A variety of forms may be in existence at any instant
in time to deal most effectively with the issues of the

moment. It can get messy, but if everybody understands how the things work and how to operate within them, then it can work fine. Of course, people need to be educated and trained how to operate in teams, task forces, and their variants.

It should be apparent that the organizational characteristics I have described for high-tech organizations, with their flexible structure and loose "permission structures," will *allow* the constitutive human agents plenty of latitude to use their uniquely human cognitive skills for adaptation and innovation, but how do we *encourage* them to, and how do we make their efforts coherent, so that chaos and disorder will not result? When we humans are properly challenged and motivated, we love solving problems and coming up with new ideas. On the other hand, we are entwined in many relationships other than the job in our complex society and we have a limited attention span, so we tend to fall into the habit of doing just enough to get by in some of our relationships in order to focus our creative energies on the more interesting or challenging ones. So the art of high-tech management is quite simple to state—do not let the organization believe "business as usual" is good enough to get by.

High-tech managers know that in order to succeed the organization must always be prepared to cope with changes in its external environment, and they know that the nature of external change is relatively long periods of slow change followed by short periods of very rapid change (punctuated equilibrium). They also know that there is no "one best way" to do things; the capacity of human cognition to adapt and learn is essentially unbounded, and the inexorable advance of technology continually offers new possibilities. Further, they know that creative change can come in two flavors, I'll call them adaptation and innovation. Adaptation is incremental improvement by continually trying small changes in an activity or process, keeping those that work. Innovation is dramatic improvement by seeing different ways to approach the problem. Adaptive processes are low-risk, low-return per step, but over time lead to major returns through compounding, while innovation is high-risk, high return per step. So high-tech managers push their organizations to continually experiment with new ways to do things, blending both adaptive and innovative efforts. If there is no external threat or opportunity to focus on at the moment for a particular part of the organization, then focus on continually improving the quality and efficiency of current activities.[2] And of course the best way to succeed is, rather than to react to the changes in environment, to create by your own innovations those changes which will be viewed by your competitors as problematic changes in their external environment.

In an organization which demands constant experimentation, it is essential to realize that most experiments fail, but the ones that succeed more than

---

[2] Of the plethora of management "fads," the one that I believe best explicates the principle of continuous improvement, can be applied to all functions of all organizations, and will stand the test of time, is Total Quality Management (TQM) and its variations.

make up for the costs of the failures. So the organizational incentive and reward systems (both financial and psychological) must reflect this; success should be handsomely rewarded, but most importantly, failure should not be punished. Only failure to experiment should be punished. The attitude toward failure should be "that didn't work as we had hoped; what have we learned from that, and what shall we try next?"

To those who are steeped in the old paradigm of organizations as machines to be designed, and managers as "controllers" of the machine, it might seem that the kinds of organizations I've described above cannot achieve sufficient coherent, coordinated action to carry out their purpose. Surely allowing people to constantly experiment and change things, not to mention having the latitude to sometimes act to further their selfish personal objectives over those of the organization, must result in chaos. How do you control such an organization? What is the "glue" that hold things together? The answer is easily understood when organizations are viewed from a complex adaptive system perspective.

Humans, as a consequence of our evolutionary history, are naturally inclined to cooperative activities. We could not have survived as a species otherwise. And our capacity for self-organization is obvious everywhere; John Holland gives the example of New York city: "New Yorkers of all kinds consume vast stocks of food of all kinds, with hardly a worry about continued supply.. yet [the city] has no central planning commission that solves the problem of purchasing and

distributing supplies, nor does it maintain large re-
serves to buffer fluctuations; their food would last less
than a week or two if the daily arrivals were cut off."[7]
All that is required for any human organization to function coherently is a shared understanding of purpose
and incentives sufficient to convince its members that
their own best interest is served by orienting their behavior toward the purpose of the organization. So the
glue that holds high-tech organizations together is a
clearly communicated sense of purpose or mission,
as well as a clearly communicated and constantly reinforced set of values governing behavior, together
with incentive systems such as profit sharing and stock
options to orient collective behavior towards accom-
plishment of the purpose.

So far I have been describing high-tech organizations
from an inward-looking perspective—their organizational structures and management practices. Equally
important is the manner in which they approach their
relationships with the external world. They pay close
attention to their interactions with external organizations—customers, suppliers, competitors—and think
hard about the changes they see. They especially go
to great lengths to involve their customers in determining features of new products. In light of our four
properties of complex adaptive systems, it is easy to
see that these characteristics would be essential for
survival. An organization must recognize that it is not
only a complex adaptive system itself, but that it is
also a member of a higher-order complex adaptive
system comprising itself and the other firms with which
it interacts. Evolution, co-evolution, and punctuated

equilibrium mean the company's world is not fixed, but constantly changing, and not only that, but the exact nature of changes in behavior of other agents and introduction of new agents is not only unpredictable, but *unknowable* [8]. In the next section, I will argue that one of the most effective ways for an organization to come to understand its world as it changes is through especially productive relationships called generative relationships. When viewed this way, two trends in high-tech behavior that go against the prescriptions of classical economics can be understood.

First, the nature of business contracts is changing. The classical economics approach leads to the view that contracts should attempt to envision all possible future eventualities, and specify *a priori* the rights of each party in each case. This leads to interminable arguments and negotiations and lost time. But if the detailed nature of outcomes is not only unpredictable but unknowable, and time is the scarce commodity, why bother? After a long period in which they trended towards increasing sophistication and complexity, contractual arrangements have become more simple, and are based much more on trust. Rather than becoming obsessed with trying to make sure that a contract covers all the bases and protects them against every eventuality, however unlikely, successful organizations take the attitude that things will be worked out as situations arise. The emphasis in this environment is to stop wasting time and get on with the business at hand.

How can an organization be responsive and keep pace if it is worrying about and spending time on contract details with low probabilities of relevance?

A second trend is toward relationships with suppliers. While classical microeconomics would predict that firms would buy only from the lowest bidder with no loyalty, high-tech firms (and now many non high-tech firms) are doing just the opposite. Instead of playing off numerous suppliers against each other, these firms are reducing the number of suppliers but forming much closer relationships with the selected set. In a three year period, Motorola reduced the number of its suppliers by 70% [9]. Reallocating relationship management efforts to fewer more intensive partners rather than many arms-length partners has several advantages, such as lower transaction costs, but a crucial one is the ability better understand and adapt to changes through collective discourse and joint action.

Do the lessons of the evolution of high-tech organizations have any applicability to other sectors, such as the military or government institutions in general? I am not sure but I think probably so, for a couple of reasons. First, on the metaphorical level, both the public and private sectors deal with complex adaptive systems and organizations; people who are working together to accomplish some purpose. We also know that most of the creativity and innovation in human activities comes from cross-domain analogies. That is, you develop a deep understanding of patterns of cause and effect in one domain of experience, perhaps physics or chemistry; you see patterns in another domain that at an abstract level resemble those of the first domain, so by analogy you hypothesize about cause and effect in the second domain. One could hope that using the experience of the private sector in adapting to the rapid pace of technological advances and applying it to a military organization is just such a cross-domain analogy. Second, all organizations have certain things in common. Both private sector organizations and the military need organizational structures, methods of coordination, information systems. They each have the need to recruit and train people, supply them with tools and materials, and deal with management issues, all in rapidly changing environments. Practices that are effective for these in the private sector may well be effective in the public sector.

Strategy under Complexity

In the previous section I have described some characteristics of high-tech organizations that enable them to adapt to rapid environmental change by constant experimentation and adaptation. But what about planning, in particular long-term strategic planning? Most high-tech organizations do not attempt detailed planning beyond 12-24 months, and even those plans are viewed as a guideline around which to organize and coordinate the activities of people, subject to frequent adjustment as events unfold. When it comes to longer term time horizons, they are highly skeptical of the standard methodologies of strategic planning that have been in vogue for many years, which are based on a presumption of underlying order that can be inferred. While many go through the motions of using the standard techniques, they place much more emphasis on the "gut-feel" of the key thinkers in the organization when it comes to decisions about major long-term investments and directions. In a human world that exhibits the properties of complex adaptive systems, implying unpredictable and unknowable novelty, is there any benefit to be gained by trying to think about the longer term? How should one go about it? My colleague David Lane and I [10] have developed some partial answers to these questions, and in this section I want to briefly introduce some of our ideas.

First it is useful to make some distinctions about foresight horizons; how far ahead the strategist thinks he can foresee events. Foresight horizons can be *clear, complicated,* or *complex.* To illustrate, I quote from the paper by Lane and me:

> Picture an 18th century general perched on a hill overlooking the plain on which his army will engage its adversary the next day. The day is clear and he can see all the features of the landscape on which the battle will be fought—the river and the streams that feed it, the few gentle hills, the fields and orchards. He can also see the cavalry and infantry battalions positioned where he and his opponent have placed them, and he can even count the enemy guns mounted in the distant hillsides. The battle tomorrow will consist of movements of these men across

this landscape, movements determined in part by the orders he and his staff and their opposite number issue at the beginning of the day, and in part by the thousands of little contingencies that arise when men, beasts, bullets and shells come together. While he cannot with certainty predict the outcome of all these contingencies, nor of the battle that together they will comprise, he can be reasonably sure that one of a relatively small number of scenarios he can presently envision will actually come to pass...The general's uncertainty has a clear terminal date: tomorrow, when the battle will have been fought and either won or lost...the general knows what he is uncertain about: not only which side will win the battle, but also the kinds of events that will turn out to be decisive...The general has a *clear* foresight horizon.

Now think about a U.S. cavalry column marching through an uncharted section of Montana in the early 1870s. The commanding officer cannot know the location of the nearest river or whether there will be an impassable canyon on the other side of the hills looming over his line of march. Nor does he know where the Indian tribes who inhabit this country have established their camps or whether they are disposed to fight should he come into contact with them. He knows the general direction in which he wants to take his men, but it would not pay him to envision detailed forecasts of what the next days might hold, because there are too many possibilities for unexpected things to happen. Instead, he relies on his scouts to keep him informed about what lies just beyond his own horizon, and he stays alert and ready for action. He in confident that he will recognize whatever situation he encounters, when he encounters it...The cavalry commander is concerned with getting his troops to their assigned destination, so his time horizon of relevant uncertainty is a matter of days or weeks...He could frame propositions about almost anything likely to be relevant to the completion of his mission, but it would amount to a very long list, most items of which would turn out not to matter anyway... The cavalry commander's foresight horizon is *complicated*. He know the kinds of thing that might happen , but because of the sheer number of possible geographical, meteorological and social combinations it is difficult to imagine them all at the outset of his mission. Nonetheless, he thinks he knows how to find out about the eventualities that are likely to matter in time to respond efficaciously to them.

Finally, imagine the situation of a Bosnian diplomat in early September 1995 trying to bring an end to the bloodshed in his country. It is very difficult to decide who are his friends and who his foes. First he fights against the Croats, then with them. His army struggles against an army composed of Bosnian Serbs, but his cousin and other Muslim dissidents fight alongside them. What can he expect from the UN Security Forces, from the NATO bombers, from

Western politicians, from Belgrade and
Zagreb, from Moscow? Who matters, and
what do they want? On whom can he rely,
for what? He doesn't know—and when he
thinks he does, the next day it changes. The
Bosnian diplomat has an uncertain time
horizon—there is no end in view. He would
be at a loss to name all the actors and events
that could affect the outcome of the drama
of which he is a part. In fact, no one could
name them, because in the working out of
the drama new actors keep getting drawn
in and they create new kinds of entities—
like the Rapid Deployment Force or the
abortive Moscow Peace Meeting—that
simply could not be predicted in advance.
The Bosnian diplomat's horizon is certainly
complicated, but there is more to it than that.
Unlike the cavalry commander, his problem
is not just to negotiate his way a fixed
landscape composed of familiar if presently
unknown features. The social landscape
through which he moves constantly deforms
in response to the action he and others take,
and new features, not previously envisioned
or even envisionable, emerge. Since his
destination is always temporally beyond his
current foresight horizon, the connection
between what he does and where he is
going is always tenuous and hence
ambiguous. Inhabiting as he does a world
of emergence, perpetual novelty and
ambiguity, the Bosnian diplomat's foresight
horizon is *complex*.[10]

If an agent has a clear foresight horizon, then the time-
honored methodology of Decision Analysis is

appropriate for strategic planning. Determine the set of possible strategies, assess the outcomes of each and their probabilities, evaluate the relative value of each outcome, and calculate the optimum strategy. In complicated foresight horizons, the hopelessly large number of possible outcomes and the difficulty of assessing probabilities, let alone assigning values, forces strategic planning to become the organization of processes of continuous experimentation, exploration, and rapid adaptation. This is the motivation for the recent spate of literature about 'the learning organization' [11,12]. But in complex horizons the very structure of the world in which the agent exists is undergoing change. What does strategy mean when "your world is under active construction, you are part of the construction crew, and there is not any blueprint"?[10]

Complex foresight horizons emerge when cascades of change occur in agents, artifacts, and their relationships. These changes have two dimensions: cognitive and structural. By cognitive change we mean changes in interpretation by human agents of their world; who the other agents are and what they do, what artifacts there are and what their function and value is, and what agents interact in what ways with which other agents and with what artifacts. By structural change we mean the emergence of new types and instances of agents and artifacts (and the disappearance of others), coupled with new and rearranged relationships between agents and artifacts. These two dimensions are coupled by *reciprocal causality*—cognitive reinterpretations of the world lead to new actions by agents which lead to new relationships with other agents and artifacts; and structural changes observed and experienced by agents lead to new interpretations of their world. Thus we have a dynamic feedback loop, and we know that feedback loops can be stable (negative feedback) or unstable (positive feedback). In our context, instability means *constructive* positive feedback, the emergence of new entities and relationships, resulting in complex foresight horizons.

Although human agents can passively observe aspects of their world with which they do not directly interact and make interpretations, the most important stimulation to reinterpretation comes through *action*, in particular *interaction* with other agents. Every agent engages in relationships—recurring patterns of interaction—with a relatively small number of other agents, and it is through these relationships that the agent can learn best about its world and changes to it. Most relationships—for example, impersonal buy-sell market interactions—do not permit the kind of information exchange that can stimulate innovative reinterpretations of the world by the participants. But a few relationships—Lane and I call them *generative relationships*[3]—do stimulate cognitive reinterpretations of the world by their participants, leading to the cascades of change of constructive positive feedback. So the dynamic feedback process that generates complex foresight horizons goes like this: generative relationships induce cognitive reinterpretations of the world

---

[3] for extended discussion of generative relationships, see [8] and [10].

which lead to actions which cause structural change which generates possibilities for new generative relationships.

To illustrate the dynamics of generative relationships, I can cite an example from my experience in building ROLM Corporation. After six years in the mil-spec minicomputer market, we diversified into the telephone PBX market in 1975. This was a billion-dollar market dominated by AT&T which had been stable for a long time. The other participants in this market, all large companies, had long-established presence and market shares that had been relatively stable for decades. But two things had happened to destabilize the status quo. First, digital technology for switching and control was evolving very rapidly but these complacent competitors continued to use old electro-mechanical switching and control technology in their products. Second, the industry had become deregulated by the Carterphone decision in 1968, allowing PBXs to be marketed competitively, rather than available only through the local telephone service monopoly. By 1974 nothing much had happened; it was still a billion-dollar market dominated by AT&T. ROLM developed a digital, computer-controlled PBX which turned out to be wildly successful. While there were no doubt many contributing factors to our success, one of the most interesting involves the changes over time in the perceptions we and our customers held about the artifact and our relationship to it. These changes were fundamental to the co-evolution of the market, the players, and the technology.

The advanced technology introduced in the ROLM PBX could be considered analogous to the biological evolution of the nervous system. While it initially provided new useful functions, it also provided a flexible platform for further evolution of radically new functions. In the biological sphere, the evolution of the nervous system to the human brain is measured in millions of years, while in the time frame of functional evolution of technology in the human world is measured in years or even months. In the initial version of the ROLM PBX, we programmed the embedded control computer with all the functions we thought could be useful to organizations, such as least-cost routing of long-distance calls, automatic dialing, and call detail recording. We knew there might be other functions that would turn out to be useful, but we had no idea what they might be. ROLM focused on telecommunications managers of the very largest companies as a key market segment. We did that because these large firms were very sophisticated with large telecommunication budgets and centralized decision making, and the new functions of our product had greater relative benefit for them than for smaller companies. It was initially very hard to make inroads with these individuals, because they were used to buying whatever AT&T told them to (a situation very similar for early innovators in the computer industry who had to compete with IBM). But we felt that if we focused intensely on serving these customers we could convince them. A few tried our product and found that not only did it do what we said it would do, but they saved so much money that they became heroes in their own companies. But more importantly they began to relate to us other needs that they had. They would come back and say, "We've been thinking of buying this automated call distribution system from Collins, but we only have fifty people handling incoming calls to our service department, whereas the Collins system is designed for thousands of airline reservation agents and is uneconomical for us; why couldn't you program these kinds of features into your PBX?" We asked our engineers how hard that would be to do, and realized it would be fairly easy to do. We went around to some other customers and explained the application, and it turned out almost everyone of them had had very similar needs. So within a year we incorporated an Automatic Call Distribution function in the next version of the product, and it was very successful. And other ideas began to emerge from our customers, such as centralized attendant service, that drove the continued transformation of the product. The results of these intense working relationships between manufacturer and consumer not only evolved the nature of the product, they also transformed our company and the whole PBX industry.

As a result of these interactions, we changed our idea of what ROLM was all about. We were not developing telephone systems, we were developing line-of-business communication systems for reducing costs and increasing the efficiency of organizations. With that new mindset, all kinds of new possibilities opened up about new applications of our technology. And as we introduced a steady stream of new innovations every few months, we continued to distance ourselves from the old-line competitors, who were accustomed to product cycles of many years.

The telecommunication managers who were early adopters of the ROLM PBX enjoyed transformations as well. Because of the benefits they delivered by embracing the new technology, they gained credibility and promotions within their companies. They previously had a relatively low level position on the corporate ladder—much lower than the MIS manager—because with the old technologies there wasn't much possibility of innovation. Their promotions began to put them on a par with MIS managers. At the annual meetings of the professional association comprising their peers —the International Communications Association— they would give formal presentations about the productivity-enhancing capabilities of the ROLM PBX, and later over drinks in the bar describe to their peers the personal rewards and recognition they had won. This led to a surge of interest by other large companies, which then stimulated interest by smaller companies who look to the larger companies for leadership. The rapidly increasing revenues to ROLM in turn allowed an even higher level of investment in continuing product innovation, and this virtuous cycle of "increasing returns" [13] allowed ROLM to emerge as a major force in a transformed industry.

In a span of five years, an unknown company, ROLM, had captured the second largest market share in a market that had been stable for decades. By 1980, three companies—AT&T, ROLM, and Northern

Telecom—had 80% of the U.S. PBX market. All of the other original major PBX manufacturers had been eliminated or marginalized, and a handful of new players had footholds. Interestingly, the same three entities (ROLM is now owned by Siemens) continue to dominate the market in 1996, sixteen years later. This provides a good example of punctuated equilibrium; the PBX market was stable for many years, then underwent a transition over only 5 years to its present stable state. I believe a key reason for ROLM's success was developing generative relationships with its key customers, leading to positive feedbacks that accelerated its rate of product innovation and market acceptance.

If we interpret the ROLM story using the abstract terms of the dynamics generating complex foresight horizons, it goes like this. A small agent (ROLM), looking for new opportunities, sees a possibility of using an artifact about which it has deep knowledge—small computers—as the basis for making an improved version of another artifact—a telephone switching system (PBX). After developing the new artifact, the company must form new seller-buyer relationships with a class of unfamiliar agents—large companies with significant telecommunication costs. After forming a few such relationships, some of the relationships become generative. The telecommunication managers of the large companies, having demonstrated the hoped-for large cost savings with the new PBX, receive unaccustomed accolades from their organizations, and realize that the possibility exists to continue to beneficially transform their own identity in the organization by additional applications of the new artifact. They turn to ROLM with requests for enhancements to the PBX to enable the new applications. This leads ROLM to realize that the possible functionality of the artifact it has designed is much broader than just traditional PBX features, implying a much larger market, and it focuses its key engineering talent to pursue these ideas. ROLM reinterprets its mission (identity) as providing business communication systems, not just telephone systems. At the same time, the successes of the early customers spread via their professional relationships with peers in other companies, leading to an exponential increase in new agent relationships for ROLM (some of which also generate new ideas), providing rapid increase in revenue, which in turn allows increased investment in product enhancements. This virtuous circle leads to explosive growth for ROLM and rapid capture of market share. So we see that the generative relationships led to reinterpretation of self-identity by both ROLM and the telecommunication managers, as well as reinterpretation of the functionality of the new artifact, and these in turn led to structural change (dramatic shifts in market share) in what had been a stable market, as well as major changes in the perception of what functionality constituted a modern business voice communication system.

But why didn't other old-line players react quickly to preserve their position, and why didn't other computer-knowledgeable companies with superior resource bases muscle their way into this newly energized market? I believe the answer is that *in order to survive and prosper during cascades of change, an organization must: first, be embedded in the generative relationships that cause the changes, and second, be capable of focused, rapid action in response to perceived opportunities.* If an agent is in a position to comprehend change only by observing the end structural results rather than the earlier cognitive shifts that led to the structural results, it will have great difficulty moving rapidly enough to succeed. And if those agents who are in the generative relationships do not exploit the opportunities quickly, they are at risk of eventually being displaced by those with larger resources. Although the old-line PBX competitors had existing relationships with their customers, these relationships did not become generative for two reasons, size and complacency; lulled into a false sense of security by years of "business as usual," they did not feel a need to maintain continual intense discourse with their customers, and when they belatedly realized the implications of computer-controlled PBXs, they were too big and bureaucratic to respond quickly enough. Similarly, by the time potential new competitors outside the industry recognized the structural changes taking place, it was too late to insert themselves in an effective way.

Strategic Practices

The foregoing discussion and story provide the basis for a partial answer to the question of what strategic thinking means when an organization finds itself with a complex foresight horizon. Lane and I [10] suggest that such organizations should put into place two strategic practices: *populating the world*, and *fostering generative relationships*. Populating the world is a process of discourse to construct and interpret a representation of the external environment—who and what are the agents and artifacts that constitute the world, what are their relationships, and how are they changing? This entails, of course, gathering information from many sources, but most importantly, pattern recognition and interpretation. Fostering generative relationships is an attempt to secure a position in the world which will enable the organization to recognize and influence emergent opportunities. Based on the organization's current interpretation of its world, it invests resources in existing relationships that have the potential for—or already demonstrate—generativeness, and it seeks to establish potentially generative relationships with new agents.

If it is true that generative relationships are an important aspect of success in complex horizons, then how are they fostered? After all, I have argued that their benefits are unforeseeable and that not all relationships become generative. The generative potential of a relationship can be analyzed by assessing the degree to which the following essential preconditions are met:

There must be *aligned directedness*. This simply means the participants have a compatible orientation of their activities; for example, one party is interested in using an artifact, the other in supplying it. Or two nations are concerned about defending themselves from a common potential aggressor. Or the Army and Navy are each trying to develop weapon systems on limited budgets.

Second is *heterogeneity;* the participants have to differ in key respects. They have to have different competencies, different access to other agents or artifacts in the world, or different points of view about how to think about agents or artifacts. In a sense they need be an interdisciplinary team. An example is the Santa Fe Institute's Business Network, with some thirty members from business, government, and military. They meet with the scientists, two or three times a year, in order to get exposure to new ideas. They are gathered around a common set of ideas and metaphors about complex systems and a number of novel joint projects have emerged. Of two nations concerned with defense, one has a strong navy, the other a strong army, and each has alliances with other nations.

Mutual directedness is needed. It is not enough to have synergistic interests and differing perspectives, but the agents must seek each other out, and develop a recurring pattern of interactions. You have to have an interactive relationship to begin with, before it can become generative. There are many kinds of natural role-based relationships, such as supplier-buyer or trading partner, and these are usually the seeds of generative relationships. Generative relationships can arise serendipitously from existing natural relationships, or an organization may seek out new relationships based on its perception of generative potential. Within an organization, management may perceive the possibility for generative potential between two sub-organizations, and create incentives for mutual directedness. For example, if a portion of the budget for new weapons systems were earmarked for common sub-systems or technology developed jointly and endorsed by all three services, it might induce new relationships that could turn out to be highly generative.

The fourth precondition for generativeness is *permissions*. The individuals interacting in the relationship have to have appropriately matched permissions or authorizations from their respective organizations to engage an open and extensive level of disclosure and dialogue. Without this, the generative potential is blocked. In relationships between organizations with multi-level reporting hierarchies, generative potential is greatly enhanced by establishing regular discourse between the responsible individuals at each hierarchical level with their peers in the other organization. This not only allows quick adjustment of mis-matched permissions and response to action opportunities, but provides even more heterogeneity in the relationship because of the differing range of perspective and knowledge inherent at the various hierarchical levels.

Finally, there must be *action opportunities*. As ideas for new possibilities arise from continued interaction, there has to be the opportunity to engage in joint action based on the ideas. Relationships that involve only talk do not last long or deeply affect agent identities. Action itself more clearly reveals the identities of the participating agents and enhances the development of mutual trust. It is interesting to consider what might have happened if the U.S. and USSR, with an aligned directedness toward strategic arms limitations, had chosen to proceed not by sending a small team of negotiators to Geneva to spend years sitting across a table talking at each other (preceded by years of arguments on the size and shape of the table), but rather by a process of taking small joint actions such as destroying a handful of weapons with mutual inspection, then another step based on the experiences of the first, and so on. Another reason for action opportunities is that new joint competences can emerge only out of joint action, and these joint competences lead to changes in agent identities and even to the emergence of new agents.

Although I have framed this discussion of generative relationships in terms of interactions between independent organizations such as companies or nations, the ideas are just as valuable applied to dependent organizations, such as departments within a company. Dramatic innovations can come about when functional sub-organizations depart from the norm of viewing their dependence relationships with other sub-organizations as a necessary evil that gets in the way of accomplishing their purpose, and instead develop discursive dialogs oriented around understanding each other's problems and initiating actions to improve the efficiency of both. One of the key responsibilities of management should be the maximization of the generative potential of relationships, both within his own (sub-)organization and with other (sub-)organizations.

Conclusion

The rapid rate of change in our modern world, driven by the enabling technology of the transistor, has strained the ability of many organizations to function effectively. One reason is that the old intellectual framework presuming a stable, or at least slowly changing, economic social order—upon which the conventional management wisdoms are based—does not apply in rapid transition periods such as we now experience. This paper has argued that applying the metaphors of the science of complex systems to the human world can provide a new intellectual framework for the management of organizations, within which the successful attitudes, methods, and practices that have evolved in the high-tech sector over several decades can be seen to make sense. High-tech organizations understand that time is the scarce commodity and people are the key asset, which has resulted in common practices: loose permission structures rather than strict operating procedures; reliance on informal and temporary organization structures rather than rigid hierarchies; incentives that reward experimentation and don't punish failure; reliance on a shared sense of mission and set of values to ensure coherence; and simple contracts and close relationships with other organizations. There is a high likelihood that at the proper level of abstraction, these practices can be applied to organizations in all sectors which face rapid change, including the military and international relations.

The prospect of unpredictable and unknowable events and emergent entities may seem to make the concept of long-term strategic planning useless. But an understanding of the mechanisms by which such changes come about—reciprocal causation between human organizations reinterpreting their world and acting accordingly, and structural change emerging from aggregate actions causing organizations to reinterpret their world—leads to practices that can allow organizations to proactively improve their prospects for success. Two such practices have been discussed: populating the world—the continual reinterpretation of the organizations, institutions, artifacts and relationships that comprise one's environment; and fostering generative relationships with selected organizations to maintain a position from which to participate in the construction of the emerging world.

Acknowledgment

I am indebted to Dr. David Alberts for valuable assistance in the preparation of this paper.

End Notes

1. P. Anderson, K. Arrow, D. Pines (editors), *The Economy as an Evolving Complex System,* Addison-Wesley (1988).

2. T. Kuhn, *The Structure of Scientific Revolutions,* University of Chicago Press, Chicago (1970).

3. Rosenau, James N. "Many Damn Things Simultaneously: Complexity Theory and World Affairs."

4. P. Drucker, *Post-Capitalist Society,* HarperCollins Publishers, New York (1993).

5. "World Economy Survey," *The Economist*, September 28, 1996.

6. "Netspeed at Netscape," *Business Week,* February 10, 1997.

7. J. Holland, *Hidden Order,* Addison-Wesley (1993).

8. D. Lane, F. Malerba, R. Maxfield, and L. Orsenigo, "Choice and Action," *Journal of Evolutionary Economics* 6, 43-76 (1996).

9. "Tying the Knot," *The Economist*, May 14, 1994.

10. D. Lane and R. Maxfield, "Strategy under Complexity: Fostering Generative Relationships," *Long Range Planning* 29 (2), 215-231 (1996).

11. P. Senge, *The Fifth Discipline: the Art and Practice of the Learning Organization,* Doubleday/Currency, New York (1990).

12. D. Garvin, "Building a learning organization," *Harvard Business Review,* July-August, 78-91 (1993).

13. B. Arthur, "Complexity in economic and financial markets," *Complexity* 1 (1), 20-25 (1995).

Command and (Out of) Control: The Military Implications of Complexity Theory

John F. Schmitt

> I shall proceed from the simple to the complex. But in war more than in any other subject we must begin by looking at the nature of the whole; for here more than elsewhere the part and the whole must always be thought of together.
>
> —Carl von Clausewitz

The greatest and most direct military implications of complexity theory are likely to be in the area of command and control. Complexity theory is command and control theory: both deal with how a widely distributed collection of numerous agents acting individually can nonetheless behave like a single, even purposeful entity. The emerging sciences suggest that war is a radically different type of phenomenon—with a different operating dynamic—than typically understood in the American military. While radically different than commonly understood, war may have much in common with other types of nonlinear dynamical systems such as, as Clausewitz suggested, commerce. If war is a dramatically different type of phenomenon than commonly understood, then the implications for the way we perform command and control may be—should be—nothing short of profound. As we learn more about the behavior of complex systems, we will likely come to view command and control in fundamentally different terms.

The Prevailing View of Command and Control

Military theorists have routinely turned to science to help understand and explain war. In the verifiable and reliable laws of the natural world they have sought analogies and explanations for the unfathomable occurrences of the battlefield. Most often military theorists have turned to physics—and more specifically to Newtonian mechanics—because it is the most established, most elegant, and most precise of the sciences and because its laws describing the movements of material bodies and the physical forces acting upon them seem to provide ready analogies for military forces engaging one another in combat.

The great Prussian military theorist-philosopher
Clausewitz was an avid amateur scientist and relied
heavily and explicitly on the physical sciences to provide metaphors for his military concepts. Two of his
greatest and most enduring concepts—friction and the
center of gravity—come straight out of the science of
the day. Of course, science for Clausewitz was
Newtonian science.

The Reigning Paradigm: Newton Rules

Not only does science provide metaphors and models for isolated military concepts, in our age it plays
an even more fundamental role: Newtonian science
provides the overarching paradigm which characterizes modern Western culture. In ways that we don't
even realize because it is internalized, our paradigm
shapes both our interpretation of the problems we face
and the solutions we generate to those problems.

The Newtonian paradigm is the product of the Scientific Revolution which began in the 16th century and
reached its crowning moment with Isaac Newton, who
gave his name to the resulting world view. The
Newtonian paradigm is the mechanistic paradigm: the
world and everything in it as a giant machine. The
preferred Newtonian metaphor is the clock: finely
tooled gears meshing smoothly and precisely, ticking
along predictably, measurably and reliably, keeping
perfect time.

The Paradigm Deeply Ingrained

The Newtonian/mechanistic paradigm is so deeply ingrained that it is even reflected in our everyday conversation. When things are going well, we say they are going "like clockwork." When our unit is performing well, we describe it as a "well oiled machine," or we say we're "hitting on all cylinders." We refer to our individual contribution by saying we're "just one cog in the machine." In the Marine Corps, for example, the common descriptor for an individual rifleman is "killing machine." And what is the Marine Corps' preferred metaphor for itself? It is the "lean, green machine."

We call military actions "operations," a term which has a strong mechanistic/procedural connotation, suggesting either a surgical procedure performed on an anesthetized patient or the systematic functioning of a piece of machinery. An operation conducted with noteworthy efficiency is referred to as a "surgical strike." Much less frequently do we refer to military actions as "evolutions"—a term which has biological connotations rather than mechanistic ones and suggests adaptation and adjustment rather than precise planning and procedure.

Newtonian War

The Western approach to war has been as heavily influenced by the Newtonian paradigm as any other field. So what is war according to the Newtonian paradigm like? Importantly, Newtonian war is deterministically predictable: given knowledge of the initial conditions and having identified the universal "laws" of combat, we should be fully able to resolve the problem and predict the results. All Newtonian systems can eventually be distilled to one simple concept: cause and effect. And in fact, just such efforts to quantify results in war have abounded, starting at least with the famous Lanchester equations and carrying through Dupuy's Quantified Judgment Model. In other words, Newtonian war is knowable: all the information which describes any situation is ultimately available, and the implications can be fully worked out. That which we cannot directly observe, we must be able to extrapolate.

Newtonian war is linear: a direct and proportional connection can be established between each cause and effect. (Here "linear" refers to the dynamical properties of a system rather than to linear formations or frontages on a battlefield.) Small causes have minor results; decisive outcomes require massive inputs. In the Newtonian view, linearity is a good thing because linear systems are tame and controllable; they do not do unexpected things. If you know a little about a linear system you know a lot, because if you know a little you can calculate the rest.

The Newtonian view of war is reductionist: we understand war by successively breaking it down into parts eventually small enough to understand and control with the expectation that this will allow us to understand and control the whole. The so-called "Principles of War," reduced to the mnemonic MOOSEMUSS to aid memorization (as if that equals understanding), are a prime example of this approach. Linear processes are amenable to such decomposition; nonlinear processes by definition are not.

The Newtonian/mechanistic view of war tends to see a military operation as a closed system not susceptible to perturbations from its surroundings. This leads toward an inward focus—on the efficient internal functioning of the military machine. If war is deterministic and if the machine is operating at peak efficiency, then victory ought to be guaranteed—without any need to consider external factors. The mechanistic view likewise leads to a focus on optimization—finding the optimal solution to any problem (which is based on the Cartesian assumption that an optimal solution exists). War comes to be seen as a one-sided problem to be solved—like an engineering problem or a mathematics problem—rather than as an interaction between two animate forces. In idealized Newtonian war, the enemy, the least controllable variable, is eliminated from the equation altogether.

Newtonian Command and Control

The natural result is a highly proceduralized or methodical approach to the conduct of military operations—war as an assembly line. Newtonian command and control tends to be highly doctrinaire—heavy on mechanistic and elaborate procedures. The mechanistic view recognizes that war may appear disorderly and confusing but is convinced that with sufficient command and control we can impose order, precision, and certainty. We can eliminate unpleasant surprises and make war go "like clockwork." Just as the Scientific Revolution sought to tame nature, the Newtonian approach to command and control—especially with the help of the information-technology revolution—seeks to tame the nature of war.

Newtonian command and control thus tends to involve precise, positive control, highly synchronized schemes and detailed, comprehensive plans and orders. Perhaps the best metaphor is a chess player moving (i.e., controlling) his chess pieces. Control measures abound, compartmentalizing the various components of the military machine and specifying how those compartments cooperate with one another. Synchronization (the timepiece metaphor applied to military operations) is merely the example nonpareil of Newtonian war: the military as one huge, highly efficient and precise machine—ticking along like a fine Swiss watch.

Newtonian command and control is microscopic command and control. Just as classical mechanics studies a system by studying the behavior of each component in the system, Newtonian command and control seeks to control the military system by positively controlling each component in the system. In military lexicon this is known as detailed control. In this setting, "command" and "control" are seen as working in the same direction: from the top of the organization toward the bottom. See figure 1. The top of the organization imposes command and control on the bottom. Commanders are "in control" of their subordinates and the situation, and subordinates are "under the control" of their commanders. The worst thing that can happen in such a system is to "lose" control.

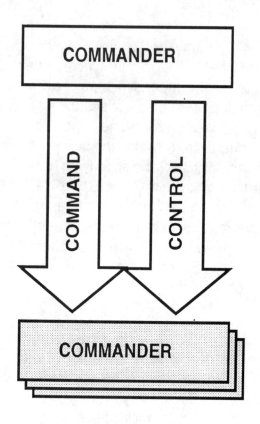

Fig. 1.

**Newtonian, or traditional, command & control—
command and control seen as unidirectional.**

The object of Newtonian command and control is to gain certainty and impose order—to be "in control." Near-perfect intelligence becomes the expectation. We pursue 95-percent certainty within a battlecube 200 miles on each side and we actually expect that we can achieve it. Consider this passage by Richard Dunn from McNair Paper No. 13:

> Increased battlefield "visibility"—provided by enhanced C3I—allows us to grasp the battle much more precisely and quickly. Thus, technology has made warfare much more certain and precise than was ever thought possible....For all intents and purposes, commanders can get a technological God's eye view of the entire battlefield.

We believe we can blow away Clausewitz's "fog of war," and if we fail to do so, it is only because our information technology is not quite capable enough yet—but we redouble our acquisition efforts and promise ourselves it will be soon.

The Problem: Reality Catches Up

The Newtonian paradigm offers a neat, clean and intellectually satisfying description of the world—and of war. There is only one problem: it does not match most of reality. When distilled to this level, the Newtonian model of war is manifestly ridiculous. When we reduce it to these terms, I think few people would argue that war is actually this way. And yet, much of the current American approach to command and control is based precisely on the unquestioned assumption of this model. Futurist Alvin Toffler states that while some parts of the universe may operate like machines, these are closed systems, and closed systems, at best, form only a small part of the physical universe. Most phenomena of interest to us are, in fact, open systems, exchanging energy or matter (and, one might add, information) with their environment. Surely biological and social systems [of which war is one] are open, which means that the attempt to understand them in mechanistic terms is doomed to failure.

This suggests, moreover, that most of reality, instead of being orderly, stable, and equilibrial, is seething and bubbling with change, disorder, and process.

The Newtonian paradigm was so compelling, so neat, so logical—in short, so "right"—that it saw and imposed regularities where none existed. For the sake of finding solvable problems, science simplified reality by assuming an idealized world. It connected the discontinuities and linearized the nonlinearities—in short, it simply ignored all the countless inconsistencies and surprises that make the world—and war—such a complex and interesting problem.

The evidence is unmistakable: the Newtonian paradigm no longer satisfactorily describes most of our world (if it ever did). Science is slowly coming to recognize that the world is not remotely an orderly, linear place after all. We need a new paradigm, and once again science may provide the catalyst. It is not after all a Newtonian battlefield: it is a nonlinear dynamical battlefield.

The Emerging View: Nonlinear Dynamical War

So what is war if not a classical Newtonian system? War is fundamentally a far-from-equilibrium, open, distributed, nonlinear dynamical system highly sensitive to initial conditions and characterized by entropy production/dissipation and complex, continuous feedback. Rather than thinking of war as a structure at equilibrium, we should think of it as a standing wave pattern of continuously fluxing matter, energy, and information. War is more a dynamical process than a thing.

The principal law of thermodynamics—the supreme Law of Nature, in fact—is the Second Law which establishes that any natural process involves an overall increase in randomness or disorder—that is, an increase in entropy. The law of increasing entropy applies to war as much as to any other natural phenomenon. The driving force of all natural change in the universe, constructive as well as destructive, is the random and undirected dispersal of energy.

In thermodynamics, equilibrium is the uniform static state of a system in which no further heat transfer is possible. It is the state of maximum entropy. Near equilibrium, systems tend to behave in a fairly linear fashion; it is when the system is forced far from equilibrium that it becomes highly responsive to fluctuations—sensitive to initial conditions—and nonlinear behavior arises. It is here that immeasurably small influences—"countless minor incidents," Clausewitz called them—can cause the system to veer off into an unpredictably and qualitatively different behavior pattern. It is here that the Second Law actually becomes a creative force through the local dissipation of entropy by leading to the spontaneous generation of structure, complexity, and life.

As an open system—continuously exchanging matter, energy, and information with other systems and with the environment at large—war is in a continuous state of flux. It is never at equilibrium, although some manifestations of war may be nearer than others—such as the stalemate of the First World War western front, which may have been as close to thermal equilibrium as any war has ever been. War is driven away from equilibrium by influxes from its environment—in the form of physical matter (or materiel) but also in the form of leadership, political motive, training, creative tactics, or any source of energy or information which tends to inject into the system the capacity to do coherent work. War is damped according to the Second Law and its universal property of entropy—which Clausewitz called "friction"—through the attrition of men and materiel, obviously, but also through fatigue, the loss of morale, poor tactics, uninspired leadership, or any other sump which drains the system of its capacity to do coherent work. At its most fundamental war can be thought of as an exchange of matter, information, and especially energy between linked, open hierarchies. Engaging an enemy by fire can be thought of as a transfer of energy from one component to another with the intended result of increasing the entropy of the latter. These exchanges take place in a complex network of simultaneous, distributed linkages between various elements at various levels in each hierarchy. Some of these linkages are tight, some are loose. Some are direct, some are indirect. See figure 2.

Feedback is a pervasive characteristic of practically all open systems, including war. As compared to Newtonian systems, which tend to have minimal feedback mechanisms, war is characterized by a complex, hierarchical system of feedback loops, some designed

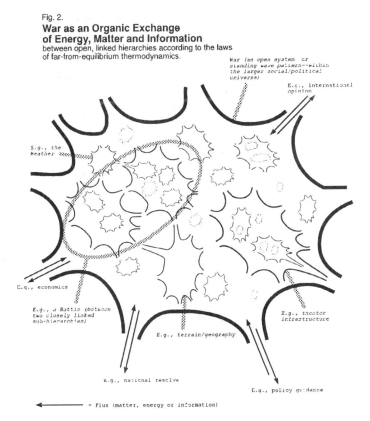

Fig. 2.
War as an Organic Exchange
of Energy, Matter and Information
between open, linked hierarchies according to the laws
of far-from-equilibrium thermodynamics.

War (an open system or
standing wave pattern—within
the larger social/political
universe)

E.g., international
opinion

E.g., the
weather

E.g., economics

E.g., a Battle (between
two closely linked
sub-hierarchies)

E.g., theater
infrastructure

E.g., terrain/geography

E.g., national resolve

E.g., policy guidance

◀———— = Flux (matter, energy or information)

but many unintended and unrecognized. Whether positive or negative, feedback results are by definition nonlinear.

War's essential dynamic comes from its being a complex, distributed system. Economic theorist F.A. Hayek coined the phrase "extended order" to describe economies driven by individual agents, but the term applies equally to war. War is an extended order: its universal nature simply cannot be captured in one place but emerges from the collective behavior of all the individual agents in the open system interacting locally in response to local conditions and partial information. In this respect, decentralization is not merely one choice of command and control: it is the basic nature of war. Centralized command and control represents an effort to muscle the system into some unnatural position—which is not to say, however, that it won't sometimes work more or less given enough energy and effort.

Information in war is, to borrow another of Hayek's phrases, "essentially dispersed." Again Hayek was writing about economics but he could just as easily have been writing about military command and control:

> This dispersed knowledge is essentially dispersed, and cannot possibly be gathered together and conveyed to an authority charged with the task of deliberately creating order.... Much of the particular information which any individual possesses can be used only to the extent to which he himself can use it in his own decisions. Nobody can

communicate to another all that he knows, because much of the information he can make use of he himself will elicit only in the process of making plans for action. Such information will be evoked as he works upon the particular task he has undertaken in the conditions in which he finds himself...Only thus can the individual find out what to look for...

The Result: War as a Complex System

According to practically any definition of the term "complexity," war qualifies as a complex phenomenon. In what could qualify as an excellent description of complexity theory, Clausewitz wrote:

> The military machine—the army and everything related to it—is basically very simple and therefore seems easy to manage. But we should bear in mind that none of its components is of one piece: each piece is composed of individuals, every one of whom retains his potential of friction. ...A battalion is made up of individuals, the least important of whom may chance to delay things or somehow make them go wrong.

Complexity theory deals with the study of systems which exhibit complex, self-organizing behavior. A complex system is any system composed of numerous parts, or agents, each of which must act individually according to its own circumstances and requirements, but which by so acting has global effects which simultaneously change the circumstances and requirements affecting all the other agents. Complex systems are based on the individual "decisions" of their numerous agents.

It is not simply the number of parts that makes a system complex (although more parts can certainly contribute to complexity): it is the way those parts interact. A machine can be complicated and consist of numerous parts, but the parts generally interact only in a designed way. This would be structural complexity. Instead, the type of complexity which most interests us is interactive complexity, by which the parts of a system interact freely in interconnected and unanticipated ways. Each agent within a complex system may itself be a complex system—as in the military, in which a company consists of several platoons and a platoon comprises several squads—creating multiple levels of complexity. But even if this is not so, even if each of the agents is fairly simple in itself, the interaction among the agents creates complexity. This is a significant contradiction of the Newtonian paradigm: simple causes can lead to complicated, disorderly behavior. ("Everything in war is simple," Clausewitz wrote, "but the simplest thing is difficult.") The result is a system which behaves in nonlinear, complicated, unpredictable and even uncontrollable ways. Each agent often affects other agents in ways that simply cannot be anticipated. With a complex system it is usually extremely difficult, if not impossible, to isolate individual causes and their effects, since the parts are all connected in a complex web. The element of chance, interacting randomly with the various agents, introduces even more complexity and disorder.

One of the defining features of complex systems is a property known as emergence in which the global behavior of the system is qualitatively different from the behavior of the parts. No amount of knowledge of the behavior of the parts would allow one to predict the behavior of the whole. Emergence can be thought of as a form of control: it allows distributed agents to group together into a meaningful higher-order system. In complex systems, structure and control thus "grow" up from the bottom; they are not imposed from the top. Reductionism simply will not work with complex systems: the very act of decomposing the system— of isolating even one component—changes the dynamics of the system. It is no longer the same system.

War is clearly a hierarchy of complex systems nested one inside another. From the largest military formation down to the individual rifleman, war consists of agents adapting to their environments—which include enemy agents—and in the process changing the environments of all the other agents.

Some of the processes in war may be deterministically predictable, some are deterministically chaotic, and some are probably purely stochastic. There are probably universals—variables or constants which show up in every mix—but no two battles, campaigns, or wars ever exhibit the same mix or system dynamic.

Even the same system may behave differently under different regimes or conditions. Under certain parameters—near equilibrium, before bifurcation—the system may actually behave in a fairly Newtonian way. Witness the Gulf War, for example, which I suggest was an unusually linear manifestation of war, in part because of low levels of interaction between the opposing sides. Under other parameters—when the system is forced farther from equilibrium—the same conflict may become very complex or even "go chaotic." The result is an infinitely complicated and continuously changing problem set that qualifies as mathematically unsolvable.

Implications

What does all this mean? We know what the command and control implications of Newtonian war are: we have been operating with them for more than a century. But if we treat war as a nonlinear dynamical system, the implications are dramatically different. These implications stem from two fundamental conclusions:

- War is fundamentally uncertain.

- War is fundamentally uncontrollable (at least given our current understanding of control).

Uncertainty A Sure Thing

Nonlinear dynamics suggests that war is uncertain in a deeply fundamental way. Uncertainty is not merely an initial environmental condition which can be reduced by gathering information. It is not that we currently lack the technology to gather enough information but will someday have the capability. Rather, uncertainty is a natural and unavoidable product of the dynamic war: action in war generates uncertainty. The only type of war about which we could achieve certainty would be a system at equilibrium, which would not be war at all.

Nonlinear dynamical systems sensitive to initial conditions are intrinsically unpredictable at the microscopic level, but the inability to accurately predict system behavior is not due to insufficient information about the system as was often assumed. Rather, unpredictability is a direct and irreducible consequence of the system's sensitivity to initial conditions and the nonlinear rules that govern its dynamics. The best we can hope for is to work out probabilities—or, as Hayek suggests, to focus on "prediction of the principle"—and even then the system will surprise us. Promises of a "God's-eye view" of the battlefield or Admiral Owens' dream of 95-percent certainty within a 200x200x200-mile battlespace are thoroughly Newtonian concepts that simply do not jibe with the nature of war as a complex phenomenon. The widespread belief that information technology will allow us to blow away the fog of war is a dangerous delusion which fails to understand the complex nature of war.

Control in War?

Complex systems like war simply cannot be controlled the way machines can. We should not think of

command and control as a coercive form of mechanistic control—the way an operator operates a
machine. The object of mechanistic command and
control is for the top of the organization to be "in control" of the bottom and for the bottom to be "under" the
control of the top. The worst thing that can happen is
for a commander to "lose" control of the situation. But
are the terrain and weather under the commander's
control? Are commanders even remotely in control of
what the enemy does? Good commanders may sometimes anticipate the enemy's actions and may even
influence the enemy's actions by seizing the initiative
and forcing the enemy to react to them. But it is a
delusion to believe that a commander can really be in
control of the enemy or the situation.

Is a kayaker paddling down a raging river really in
control of the situation? Does he control the river?
Does he really even control his own course? Or does
he try to steer his way between and around the rock
formations which spell disaster as the rapids carry him
along. For the kayaker, success—safely navigating
the river—is not a matter of push-button precision. For
the kayaker—as for the commander—it is a matter of
coping with a changing, turbulent situation. Command
in war is less the business of control than it is the
business of coping.

Complexity suggests it is a delusion to think that we
can be in control in war with any sort of certitude or
precision. Complexity further suggests the radical idea
that the object of command and control is not to
achieve control but to keep the entire organization

surfing on the edge of being "out of control" because that is where the system is most adaptive, creative, flexible, and energized.

Macroscopic Command and Control

The turbulence of modern war suggests a need for a looser form of influence—something more akin to the willing cooperation of a soccer team than to the omnipotent direction of the chess player—that provides the necessary parameters in an uncertain, disorderly, time-competitive environment without stifling the initiative of subordinates. Complexity suggests the need for macroscopic command and control. Command and control should not try to impose precise domination over details because the details are inherently uncontrollable. Rather, it should try to provide a broad, meaningful structure to the roiling complexity. Newtonian command and control is microscopic: it attempts to control the system by controlling each particle in the system. Complex war defies microscopic command and control and instead requires macroscopic command and control which "controls" the system by influencing the system parameters and boundary conditions.

Adaptive Command & Control

In a complex, open environment, command and control is fundamentally a process of continuous adaptation. The simple command and control model, the Observation-Orientation-Decision-Action cycle (or OODA loop), essentially describes a process of continuous adaptation to a changing situation. See fig. 3. We might better liken the military organization to a predatory animal—seeking information, learning

Fig. 3.
The OODA loop: Command & control as an adaptive process.

and adapting in its desire for continued survival—than to some "lean, green machine." Most military actions do not proceed with clockwork mechanics—as "operations"—but instead as "evolutions" along the "edge of chaos."

Rather than thinking of "command" and "control" both operating from the top of the organization toward the bottom, we should think of command and control as an adaptive process in which "command" is top-down guidance and "control" is bottom-up feedback. See fig. 4. All parts of the organization contribute action and feedback—"command" and "control"—in overall cooperation. Command and control is thus fundamentally an activity of reciprocal influence involving give and take among all parts, from top to bottom and side to side.

Mission Command & Control

This response to the problem leads to is what is known in military terminology as directive or mission command and control, in which control is an emergent property arising spontaneously: unity of effort is not the product of conformity imposed from above but of the spontaneous, purposeful cooperation of the distributed elements of the force. Subordinates are guided not by detailed instructions and control measures but by their understanding of the requirements of the overall mission. Commanders command with a loose rein, allowing subordinates greater freedom of action and requiring them to adapt locally to developing conditions. Mission command and control tends to be decentralized to increase tempo and adaptability. Discipline imposed from above is reinforced with self-discipline throughout the organization. Necessary close coordination is effected locally rather than managed centrally.

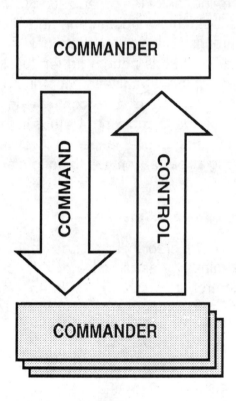

Fig. 4.

**Command and control viewed as reciprocal influence—
command as initiation of action and control as feedback.**

The critical factor in such a system is to create command parameters and other systems features which
provide the necessary guidance and level of understanding to create unity of effort without unnecessarily

constraining the activities of subordinates. In other words, how do we create the modes of agent behavior under which the necessary system control will emerge naturally? Clearly, concepts like Commander's Intent and Focus of Effort play a key role, as do the extensive education, training, and socialization of individual decision makers.

The Concept Of Evolutions?

Rather than thinking of a military action as an "operation," a predetermined plan unfolding with machinelike order and procedural precision, we should think of the action as an "evolution," a system adapting over time in response to its environment. Better yet, we should think of military action as a form of coevolution, our system evolving in response to what the enemy does and the enemy system evolving at the same time in response to us.

Complexity suggests that, just as evolution does not have a predetermined destination, military plans should not prescribe detailed end-state conditions which are instead always changing in response to developments. We should not think of a plan as a closed-form solution to a problem but as an open architecture which maximizes evolutionary opportunities. A good plan becomes the basis for adaptation through evolution. Planning is "solution by evolution" rather than "solution by engineering."

Synchronization Out Of Sync

One military command and control concept that does not mesh well with complexity theory is synchronization. Synchronization and other Newtonian models are invalidated as general operating systems. They may work moderately well within those narrow parameters under which the system behaves relatively tamely. Synchronization falls flat when faced with a complex system which does not exhibit mechanistic dynamics. In fact, healthy complex adaptive systems tend to behave asynchronously—multiple agents acting independently of one another in response to local conditions. Complexity suggests the superiority of loosely coupled, modular plans which do not rely on synchronized control for their unity of effort. Such plans allow greater latitude in execution and, importantly, are more easily modified and repaired than synchronized ones. Where synchronization occurs, it should be the result of local cooperation between agents rather than of centralized direction.

Satisfice, Don't Optimize

Complexity suggests it is rarely worth the effort trying to find the perfect plan or reach the perfect decision. It simply will not happen: there are too many interconnected variables. As geneticist John Holland has said, in a complex system "there's no point in imagining that the agents in the system can ever 'optimize' their fitness ... The most they can ever do is to change and improve themselves relative to what the other agents are doing." Instead, we should try to satisfice—find a solution that works locally and exploit the results.

Excellence Can Only Start At The Bottom

Evolution moves from the simple to the complex. Healthy complex systems evolve by chunking together healthy simpler systems. Attempts to design large, highly complex organizations from the top down rarely work, if ever. This merely confirms what successful military organizations have long recognized: success starts at the small-unit level. Build strong, adaptable squads and sections first. Train and equip them well—which includes giving them ample time to train themselves (i.e., to evolve). Give them the very best leaders. Give those leaders the freedom and responsibility to lead (i.e., let them act as independent agents). Then chunk the teams and squads together into increasingly larger units.

In Closing: Continuous Adaptation

The physical sciences have dominated our world since the days of Newton. Moreover, the physical sciences have provided the mechanistic paradigm that frames our view of the nature of war. While some systems do behave mechanistically, the latest scientific discoveries tell us that most things in our world do not function this way at all. The mechanistic paradigm no longer adequately describes our world—or our wars. Complex systems—including military organizations, military evolutions, and war—most definitely do not behave mechanistically. Enter complexity.

Complexity encourages us to consider war in different terms which in turn point to a different approach to the command and control of military action. It will be an approach that does not expect or pursue certainty or precise control but is able to function despite uncertainty and disorder. If there is a single unifying thread to this discussion, it is the importance of adaptation, both for success on the battlefield and for institutional survival. In any environment characterized by unpredictability, uncertainty, fluid dynamics, and rapid change, the system that can adapt best and most quickly will be the system that prevails. Complexity suggests that the single most important quality of effective command and control for the coming uncertain future will be adaptability.

Complexity Theory And Airpower: A New Paradigm for Airpower in the 21st Century

Steven M. Rinaldi

M ilitary theorists and analysts have frequently turned to the sciences for insight into the nature of warfare. For the past several centuries, classical physics provided a paradigm that was extensively used to frame theories of warfare.[1] Military theorists borrowed a number of concepts from classical or "Newtonian" physics and applied them metaphorically to operational art: friction, center of gravity, mass, and momentum, to name a few. In a very real sense, a large body of military art and theory came to rest upon analogies to principles of classical physics. The Newtonian paradigm thus provided a foundation for centuries of military thought, and continues to play a major role today.

Yet much as classical physics cannot describe many classes of natural phenomena, the Newtonian paradigm is limited in its application to warfare. Armed conflict has many facets that cannot be adequately addressed within this paradigm. As a consequence, theories of war built upon a Newtonian frame are restricted in scope and applicability. Military science is in need of a new framework that better describes the true nature of war.

During the past three decades, a number of new sciences have arisen and developed rapidly. Based upon nonlinear dynamics and far-from-equilibrium thermodynamics, these disciplines include catastrophe theory,[2] chaos theory,[3] and most recently complexity theory.[4] These theories push beyond some of the limitations of classical physics, and explore classes of phenomena outside the traditional linear realm. Recently, military theorists have applied these theories metaphorically to studies of conflict.[5] Complexity theory lends itself particularly well as a new paradigm for military science, promising to be a far more powerful framework than the traditional Newtonian paradigm. Although its application to military science has a short history, complexity theory already has yielded insights into military science that the Newtonian paradigm is incapable of providing. Indeed, modern military concepts such as OODA loops and parallel warfare that cannot be treated under the Newtonian paradigm (metaphorically or otherwise) fit very comfortably into the realm of complexity.

In this paper, we will explore the evolution of airpower theory in the context of the two paradigms. Airpower theory is undergoing a fundamental shift from the restrictive Newtonian to the broader complex paradigm. Theories in the post-World War I era were decidedly Newtonian in nature. Gradually, theorists proposed concepts that fit more nearly into the complex paradigm than the Newtonian framework. This trend appears to be accelerating today, as the concepts of complexity theory become more widely spread and studied. By and large, this shift has not been due to a conscious effort to develop airpower theory within the complex paradigm. Rather, complexity theory has been used principally to reinterpret existing tenets of airpower theory.

We will start by defining the two paradigms, focusing on their essential characteristics. Next, we will examine airpower theories from the post-World War I, World War II, and modern eras. In particular, we will observe how the underlying natures of the theories have shifted from the Newtonian to the complex framework. Finally, we will move from the theoretical to the practical, and examine an application of complexity theory to operational art.

Airpower theory has been transitioning to the complex framework for decades. We must complete the shift and employ the new framework in our operational thinking. For it is only in incorporating insights from complexity in practical applications that we fully exploit the power the new paradigm has to offer.

The Newtonian and Complexity Paradigms

The Newtonian paradigm has governed the way military theorists viewed warfare for many years. However, it suffers from a number of serious shortfalls. Its applicability as a framework for recent theories of airpower is increasingly questionable. As our understanding of the behaviors of complex adaptive systems increases, the complex framework becomes more relevant to studies of warfare than the Newtonian paradigm. The principle reason may be found in the concepts of linearity and nonlinearity: warfare is an inherently far-from-equilibrium, nonlinear phenomenon. The Newtonian paradigm rests firmly upon linear principles, whereas complexity theory embraces the nonlinear.

Linear systems played an important role in the development of science and engineering, as their behaviors are easily modeled, analyzed, and simulated. A linear system has two defining mathematical characteristics. First, it displays proportionality. If some input X to the system gives an output of Y, then multiplying the input by a constant factor A yields an output of AY. The second characteristic of linear systems is superposition. That is, if inputs X_1 and X_2 give outputs Y_1 and Y_2 respectively, then an input equal to $X_1 + X_2$ gives an output of $Y_1 + Y_2$. Systems that do not display these characteristics are called nonlinear. Importantly, linear systems of equations can be solved analytically or numerically. Given a set of linear equations and initial conditions, we can calculate the future values of the variables. Consequently, if we can describe a system by a linear mathematical model, we can determine its future states exactly from its given initial state. A large body of mathematics has grown up around linear systems and techniques for their solution.

Nevertheless, the vast majority of systems and phenomena in the real world are nonlinear. As their name implies, nonlinear systems do not display the linear characteristics of proportionality and superposition. Analytical solutions to nonlinear equations are generally the exception rather than the rule. Thus, the future states of nonlinear systems can often only be approximated. One method of approximating the behavior of nonlinear systems involves linearizing them, then employing linear systems analysis to the approximated system.[6] Unfortunately, such techniques suppress or even eliminate many of the important dynamical characteristics of nonlinear systems; for example, chaos cannot exist without nonlinearities. However, the advent of modern digital computers has brought about a revolution in the study of nonlinear systems. Computers have made it possible to simulate their rich dynamical behaviors such as chaos that might otherwise not exist in linearized approximations.

Linearity is the cornerstone of the Newtonian paradigm. This has several important ramifications for military theory.[7] First, warfare under the Newtonian paradigm is deterministically predictable, as effects are in principle calculable from their underlying causes. Given enough information about the current state of a conflict and armed with "laws" of combat, a commander should be able to precisely determine the outcome of the battle. The three-to-one rule of combat is a good example of a linear law. Determining the outcome of a war becomes a simple exercise if a sufficient amount of precise information is available, much as the future states of a linear system of equations can be exactly computed. A consequence of determinism in war, then, is the drive for greater quantities of ever-more-perfect intelligence from which the commander can make ever-more-precise predictions of the future. With intelligence and situational awareness approaching perfection, the Newtonian paradigm reduces fog and friction to a bare minimum just as chaos is banished from linear systems.

Reductionism is a second important consequence of the Newtonian paradigm. Reductionism is a methodology for solving problems. The analyst breaks the problem into its constituent pieces, solves each piece separately, then sums the results from the pieces to obtain the overall solution to the problem. This is a natural consequence of superposition. The history of warfare is replete with examples of reductionism. For example, targeting has largely been reductionist. Air planners generally break the enemy into a series of target systems, analyze each target system independently of all others to determine aimpoints, then sum the results to generate the overall air campaign.[8] Historical analyses of wars are frequently reductionist—what is the isolated, independent cause (or causes) that led to the outcome of the conflict? As linearity allows and even encourages this mindset, reductionism is a principle characteristic of Newtonian warfare.

A third consequence of the Newtonian paradigm is the view of systems as closed entities, isolated from their environments. Outside events do not influence such a system; the only dynamics are those arising from its internal workings. The analyst thus has an inward focus, with a concentration on efficiency. The emphasis on efficiency is especially noteworthy for military operations. How can the commander obtain the desired objectives with the least cost? What targets must the planner select to most efficiently and economically accomplish the objectives? Numerical measures of merit, such as body counts, tank kills, and aircraft losses become paramount in analyzing the flow of the battle and determining strategy. Isolated, closed systems are perhaps easier to analyze, as outside forces and influences are of no consequence. However, isolated systems form only a small fraction of the physical universe. With the global information explosion, conflicts that are truly isolated from the outside world are increasingly rare, if they exist at all.

Warfare in the Newtonian paradigm has several defining characteristics. Schmitt lists several of the more important ones:

•Warfare is highly procedural, with methodical approaches to the conduct of
military operations.

•Warfare is doctrinaire and rigidly structured, with checklists and procedures.

•Warfare has precise command and control, with rigid command and support
relationships.

•Warfare employs highly orchestrated or
synchronized schemes. Detailed plans
and orders are commonplace.[9]

The Newtonian paradigm creates a simplified, idealized view of warfare. It is an appealing, comfortable
framework as it offers simple means for analysis, methodical rules for planning and executing operations,
and the illusion of predicting the future given enough
information about the present. However, warfare is
intrinsically more complicated than this simplistic
framework allows. As Schmitt notes, the paradigm is
in need of a serious tune-up, or better, a complete
overhaul.

Complexity theory offers a broader, far more useful
framework for military theory. This paradigm is based
upon open, nonlinear systems in far-from-equilibrium
conditions. A complex adaptive system has several
defining characteristics.[10] First, it is composed of a
large number of interacting parts or "agents." The interactions between the agents are nonlinear.[11] The
interactions and behaviors of the agents influence the

environment in which the system exists. Changes in the environment in turn influence the agents and their interactions. The agents and environment thus continuously affect and are affected by each other. Second, the agents characteristically organize into hierarchies. Agents at one level of the hierarchy cluster to form a "super agent" at the next higher level. A bureaucracy or military organization illustrates the concept: a number of aircraft form a squadron, several squadrons form a wing, and so forth. Third, there are intercommunicating layers within the hierarchy. Agents exchange information in given levels of the hierarchy, and different levels pass information between themselves as well. Finally, the complex system has a number of disparate time and space scales. For example, military operations at the squad level are highly localized and may occur very rapidly compared to events at the corps level. Complex adaptive systems in widely varying disciplines appear to share these four characteristics.

Complex adaptive systems exhibit a number of common behaviors. The first is emergence: the interactions of agents may lead to emerging global properties that are strikingly different from the behaviors of individual agents.[12] These properties cannot be predicted from prior knowledge of the agents. The global properties in turn affect the environment that each agent "sees," influencing the agents' behaviors. A synergistic feedback loop is thus created—interactions between agents determine emerging global properties which in turn influence the agents. Consider a pilot and his wingman engaged in tactical operations during a conflict. The myriad of such tactical operations interact and define the courses of the operational and strategic levels of war. However, the characteristics of the strategic level of war cannot be extrapolated from individual tactical engagements. The strategic environment in turn shapes future tactical engagements for the pilots, thus completing the cycle. A key ramification of emergence is that reductionism does not apply to complex systems.[13] Since emergent behaviors do not arise from simple superpositions of inputs and outputs, reductionism cannot be used to analyze the behaviors of complex systems. The emergence of coherent, global behavior in a large collection of agents is one of the hallmarks of complex systems.[14]

A second fundamental behavior of complex systems is adaptive self-organization. As Kauffman notes, "contrary to our deepest intuitions, massively disordered systems can spontaneously 'crystallize' a very high degree of order."[15] This appears to be an innate property of complex systems. Self-organization arises as the system reacts and adapts to its externally imposed environment. Such order occurs in a wide variety of systems, including for example convective fluids, chemical reactions, certain animal species, and societies.[16] In particular, economic systems are subject to self-organization. The adjustments economies make under the rigors of war are manifestations of the dynamics of adaptive self-organization.

A third important behavior of complex systems is evolution at the edge of chaos. Dynamical systems occupy a "universe" composed of three regions.[17] The first is an ordered, stable region. Perturbations to the systems tend to die out rapidly, creating only local damage or changes to the system. Information does not flow readily between the agents. In the second region, chaotic behavior is the rule. Disturbances propagate rapidly throughout the system, often leading to destructive effects. The final region is the boundary between the stable and chaotic zones. Known as the complex region or the "edge of chaos," it is a phase transition zone between the stable and chaotic regions. According to Kauffman, systems poised in this boundary zone are optimized to evolve, adapt, and process information about their environments.[18] As complex systems evolve, they appear to move toward this boundary between stability and chaos, and become increasingly more complex. There is a direct parallel between the adaptations and substitutions made by economies due to wartime destruction and the concept of evolution at the edge of chaos. Social and bureaucratic structures display similar evolutionary patterns as well.[19] Metaphorically, we can envision the former Soviet Union existing in the stable region, Somalia in the chaotic, and the U.S. and Western European nations at the edge of chaos. Evolution toward the edge of chaos appears to be a natural property of complex systems.

The final key behavior of complex systems is their ability to process information.[20] The systems sense their environments and collect information about surrounding conditions. They then respond to this information via a set of internal models that guide their actions. Systems may also encode data about new situations for use at a later date. This characteristic is closely related to adaptation near the edge of chaos. As we shall discuss in more detail below, Boyd's OODA loop relies upon just such information processing.[21] Information processing is a common characteristic of complex systems, and enables them to adapt to changing environments.

Complexity theory provides a powerful framework for analyzing military art and science. Indeed, warfare is a nonlinear, complex, adaptive phenomenon with two or more coevolving competitors. The actions of every agent in the conflict, from individual pilots and infantrymen to numbered air forces and corps, influence and shape the environment. Environmental changes in turn cause adaptations in all hierarchical levels of the warring parties. As both allied and enemy actions influence the environment, warfare involves the co-evolution of all involved parties. When viewed from this vantage point, it is clear that the Newtonian paradigm is too limited to adequately cover the many aspects of armed conflict. Warfare is more appropriately analyzed under the complex framework than the Newtonian. It is time to shift paradigms.

Airpower Theories and Shifting Paradigms

Airpower theory is transitioning from the Newtonian to the complex paradigm. In the early years of military aviation, airpower theories fit entirely within the

Newtonian paradigm. Aspects from complexity first appeared in American airpower doctrine in the 1930s. Planners incorporated these aspects in the air campaigns of World War II. Only during the past few years have any works appeared that specifically referred to or were based upon complexity theory. For the most part, this shift has not been a deliberate attempt to incorporate principles from complexity into airpower theory. Rather, airpower theorists have increasingly proposed ideas that fall more naturally under the complex paradigm than the Newtonian paradigm. We will trace this shift by examining airpower theories from the post-World War I, World War II, and modern eras.

Airpower Theory After World War I

The Great War provided the first major test of the new aerial weapon. Following the war, proponents of airpower began to publish their theories. Two of the most prominent early theorists were the Italian General Giulio Douhet and the American General William Mitchell.

Douhet was an outspoken proponent of the aerial arm. He viewed the airplane as an inherently offensive weapon with a strategic mission. According to Douhet, the principle objectives of aerial warfare were to first obtain command of the air, then direct offensives against surface targets to crush the material and moral resistance of the enemy.[22] The airplane was unique in its ability to perform these missions. It could leap over fortified lines of defense, mass anywhere, and attack any objective in enemy territory. Consequently, the boundaries of future wars would be national boundaries, with civilians and military alike subjected to the effects of war. Douhet theorized that under a steady rain of bombs, the enemy civilian population would rise up in terror and force its government to sue for peace. He stated:

> A complete breakdown of the social structure cannot but take place in a country subjected to this kind of merciless pounding from the air. The time would soon come when, to put an end to horror and suffering, the people themselves, driven by the instinct of self-preservation, would rise up and demand an end to the war—this before their army and navy had time to mobilize at all![23]

The prerequisite for victory was to secure command of the air. He flatly stated that "to have command of the air is to have victory" and that "the command of the air is a necessary and sufficient condition of victory."[24]

Douhet was convinced that an independent air force was essential for national defense. Starting from the "axiom" that command of the air meant victory, he arrived at the conclusion that national defense could only be assured by a sufficiently powerful air force independent of the army and navy.[25] Defense against enemy airpower was futile to Douhet, as enemy forces could fly virtually undetected to and mass over any target of their choosing. Possession of a more potent offense, capable of rendering greater destruction upon the enemy, was the only acceptable solution.

Consequently, Douhet felt that the independent air force should be composed of the greatest number of bombers possible. He disdained auxiliary uses of airpower, arguing that it was "worthless, superfluous, harmful...consequently, *aerial means set aside for auxiliary aviation are means diverted from their essential purpose, and worthless if that purpose is not pursued*."[26] With command of the air equated to victory, the independent air force had to be designed to ensure this objective could be obtained.

In the United States, Mitchell was probably the most ardent early proponent of airpower. He recognized that the airplane brought a revolutionary new capability to militaries that would fundamentally change the nature of future campaigns:

> The advent of airpower which can go straight to the vital centers and entirely neutralize or destroy them has put a completely new complexion on the old system of making war. It is now realized that the hostile main army in the field is a false objective and the real objectives are the vital centers. The old theory, that victory meant the destruction of the hostile main army, is untenable.[27]

Mitchell envisioned total warfare that would affect all the citizens of a nation as the way of the future. The vital centers included civilian objectives (cities, areas where food and supplies were produced, transportation)[28] and the hostile nation's means of making war (aircraft factories, flight training schools, war materiel manufacturers, means of communications, fuel and oil production).[29] Like Douhet, he believed that an air force would need control of the air. The battle for control of the air would be the prelude to any land or sea engagements. If the United States were to lose control of its airspace in a conflict, then the enemy would be able to dictate the terms of peace "at any place within the United States that he may desire."[30]

Mitchell devoted considerable thought to the structure of the nation's future military services. The country needed a separate air force with a centralized command system to control all aspects of the employment of aircraft. The air force would have to organize its resources so that it could swiftly mobilize in event of war. This would allow the air force to strike first at any potential enemy, thus gaining a considerable strategic advantage. The army and navy would assume secondary roles. He particularly downplayed the role of the surface navy. Its large infrastructure, high cost, and the vulnerability of ships to aerial bombardment fueled his conviction that surface navies were rapidly losing their importance to national defense. Mitchell could only see prominent national defense roles for the air force, the army, and the submarine corps.[31]

Despite their visionary theories for airpower, Douhet and Mitchell were firmly grounded in the Newtonian paradigm. Both men grasped the revolutionary new capabilities of the airplane and saw the implications for the future: the ability to overfly surface forces and mass above any enemy objective, strategic attack, and total warfare. But they incorporated their airpower theories into the existing linear framework, thus expanding the body of military theory without discarding the Newtonian paradigm. In particular, Douhet exhibited a striking sense of linearity in his logic and supporting examples.

Douhet relied heavily upon linear calculations and arguments to support three important propositions. First, World War I demonstrated the defensive nature of firearms and trench warfare. Douhet asserted that improvements in firearms would favor the defense. Starting from a simplistic calculation of the number of infantrymen required to storm a trench for a given rate of defensive fire, he concluded:

> With this increased power of firearms, the offensive must, in order to win, upset this equilibrium by a preponderance of forces... But to say that the increased power of new weapons favors the defensive is not to question the indisputable principle that wars can be won only by offensive action. It means simply that, by virtue of increased fire power, offensive operations demand a much larger force proportionately than defensive ones.[32]

Increases in defensive power though improved firearms required proportionally larger offensive forces for victory. In effect, Douhet employed linear mathematics and logic to demonstrate the defensive nature of rifles, machine guns, and artillery.

His second proposition that rested solidly upon linear mathematics was the futility of defense against aerial

attack.[33] Given N targets worth defending and an enemy air force with an offensive power of X, Douhet
argued that a nation required a defensive air force of
power NX to ward off an enemy attack. Following this
logic, if a nation had twenty defensive positions requiring protection, it could only ensure its defense with
twenty times the total number of enemy aircraft. He
concluded that defense against aerial attack was absurd as a nation would be forced to tie up an enormous
amount of resources in defensive forces alone. To
Douhet, it was far less expensive and more prudent
to focus on the offensive mission of the airplane.

A third important proposition of Douhet's theory was
the inherently offensive nature of the airplane. Again,
Douhet turned to mathematics to prove his case. Starting from an explosive with a given destructive power,
he used simple, linear calculations to determine the
size of a bombing force required to completely destroy a target of 500 meters diameter. From this result,
he extrapolated the offensive potential of an air force
based on the number of aircraft in its inventory.[34] His
case for the offensive nature of the airplane was idealized and mechanically precise. But in a fundamental
oversight, Douhet's linear calculations ignored fog, friction, and chance—the nonlinear, chaotic phenomena
that are inherent to any combat environment. His arguments overlooked the myriad of factors that combine
to make aerial warfare results mathematically incal-
culable, such as imprecise bombing runs and
accuracy, mechanical failures, weather, and morale.

Warplanes are indeed offensive systems, but Douhet's supporting logic for this proposition was thoroughly linear.

Douhet and Mitchell were two important airpower theorists of the post-World War I era. Both saw the airplane as a strategic, offensive weapon system that would profoundly change the nature of future conflicts. Nevertheless, their airpower theories were rooted in the Newtonian paradigm. In essence, both men proposed extensions and modifications to the existing body of military theory. But they remained within the bounds of the Newtonian paradigm and did not incorporate any ideas from the yet-undefined science of complexity.[35] To the degree that Douhet and Mitchell represent the early school of airpower thought, we can assert that the initial theories of airpower were decidedly Newtonian in character.

Airpower Theory and World War II

American airpower theory began its transition from the Newtonian to the complex paradigm several years before World War II. We can trace the transition to the Air Corps Tactical School (ACTS). In particular, the "industrial web" theory developed at ACTS contained some characteristics of complex adaptive systems. ACTS doctrines profoundly influenced the air campaigns of World War II. As a result, concepts from complexity theory turned up in the planning and execution of the air campaigns during the war.

A large body of doctrine developed at ACTS was based upon the industrial web theory. The ACTS

instructors viewed an economy as a number of interlocked, interdependent sectors. For example, the
aluminum industry required electricity to produce aluminum from bauxite ore, and the steel industry
received raw materials and shipped finished products
via the transportation network. The linkages and dependencies created a web-like structure within the
economy. In particular, the instructors emphasized
that dependencies would often give rise to bottlenecks—specific parts of the web that were essential
for the normal functioning of the economy. Bottlenecks
had several characteristics: they were critical for the
operation of other industries, their destruction could
cause the complete collapse of or work stoppage in
an industry, and they were generally difficult to replace
or repair. The ACTS instructors considered bottlenecks to be ideal targets in the industrial web context.
Importantly, many parts of the web essential for manufacturing war materiel were also tightly connected to
civilian use. The instructors singled out electricity and
transportation as two economic sectors that were required by everyday life, civil manufacturing, and war
materiel production.

Under the stress of war, the instructors believed that
the industrial web would be severely strained and
highly vulnerable to attack. Disruption or destruction
of parts of the web would cause it to unravel, with an
impact on both the social welfare and war materiel
manufacturing capabilities of the nation:

> ...modern warfare places an enormous load
> upon the economic system of a nation,

which increases its sensitivity to attack manifold. Certainly a breakdown in any part of this complex interlocked organization, must seriously influence the conduct of war by that nation, and greatly interfere with the social welfare and morale of its nationals.[36]

Disruption of the enemy's web was therefore a primary objective in war:

> Hence it is maintained that modern industrial nations are susceptible to defeat by interruption of this web, which is built to permit the dependence of one section upon many or all other sections, and further that this interruption is the primary objective for an air force. It is possible that the moral collapse brought about by the breaking of this closely knit web will be sufficient, but closely connected therewith is the industrial fabric which is absolutely essential to modern war.[37]

The ACTS doctrine maintained that the destruction of the war materiel manufacturing base and the concomitant breakdown of the morale of enemy civilian population would cause the enemy to capitulate.

The industrial web concept incorporated two principle characteristics of complex adaptive systems. First, the web consisted of a large number of interacting, interdependent parts. However, many of the key attributes of the linkages were ignored or only superficially examined. The lectures discussed and gave many examples of first-order linkages, the direct ties between two agents in an economy. Generally, higher order linkages (indirect ties between two agents that use other agents as intermediaries) were not discussed, although they could be inferred from some of the examples. The lectures only obliquely referred to nonlinear linkages, and never by name. For example, one lecture made the points:

> •"An offensive against the various sources of energy would have a general effect upon the entire economic fabric of a nation. The amount of destruction required would be small in comparison with the magnitude of the results."

> •"The offsets of an offensive against the transportation facilities are similar to those obtained in the energy field, as the influence of the attack would be general in its disorganization power upon the entire economic fabric."

> •"Air power could thus defeat a nation by depriving it of just one commodity, steel, because no nation can successfully wage war without it."

> •"Interference with the great financial centers of a nation may produce a grave effect upon the whole nation."[38]

The instructors recognized that small inputs could lead to disproportionately large outputs. Yet it is not clear that they deeply understood or comprehensively analyzed the nonlinear nature of the linkages. Finally, the lectures rarely pointed out feedback loops, branching and serial processes, and the time-dependent nature of the linkages. In short, the ACTS theory recognized the existence of linkages, but did not fully appreciate neither the important characteristics of the linkages nor the types of complex adaptive behaviors the linkages could create.

Second, the ACTS faculty understood that economies were composed of hierarchical structures. At the lowest level of the hierarchy, different economic sectors were linked and interdependent. For example, certain manufacturing operations relied upon electrical power, and the food distribution system required the transportation network. Similarly, economic sectors were connected and interdependent across different geographical regions. Moving up to the national level, the instructors contended that a nation's economic system was tightly interlaced with its social, political, and military structures. The dependencies became absolute during a war, and disturbances in any sector would have impacts of varying intensity in the other sectors.[39] However, the hierarchical structure did not stop at national boundaries. Nations themselves were economically intertwined, with trade and money providing the linkages.[40] The spread of the Great Depression illustrated the international interconnections. The ACTS instructors clearly appreciated the hierarchical, interconnected nature of economies and national structures.

Despite this appreciation, the ACTS faculty did not fully understand the complex, adaptive nature of economies nor its implications. Indeed, aspects of the industrial web theory were thoroughly Newtonian. First, the ACTS lectures repeatedly stressed the importance of efficiency. For example, while examining the principle of the objective, one lecture noted:

> We must concern ourselves with the total efficiency of military operations, because that is our responsibility. . .The objective of the air force must contribute its maximum to attaining the national aim. Military effort of all sorts has as its ultimate aim the destruction of the enemy's material and moral means of resistance. The method of obtaining this result is not important in itself but every effort should be directed toward that result in the most efficient manner.[41]

The drive for efficiency led to the search for bottlenecks. This emphasis carried the risk of creating checklists of "panacea" target sets—minimal lists of targets perceived to most efficiently accomplish the mission. However, *efficient* target sets are not necessarily *effective* target sets, especially when an economy behaves as a complex adaptive structure rather than a Newtonian machine. Some of the experiences in World War II (especially in light of German workarounds and repair efforts) would bear this out.

A second Newtonian aspect of the ACTS theory was the presumption that the effects of bombing would accumulate. While discussing the principle of mass, one lecture stated:

> But how is air "mass" obtained? In one of two ways—either by the employment of numerically strong formations, one blow of which will destroy the objective, or by repeated blows by a relatively weaker formation, the net results of which will accomplish the desired results. Here again, is a striking difference between air and ground action . . . <u>The significant point is: in air force action results are of sufficient permanency that they accumulate</u>. So the application of the principle of mass to air force operations is a matter of providing sufficient force to accomplish each assigned mission, a sufficient rapidity of missions to prevent adequate repair between times, and sufficient missions all contributing toward the desired end.

> ...the effect of the action of a striking force against normal, permanent ground installations is cumulative, and while it may take a group of bombardment to destroy a particular objective, that same objective may be destroyed by a squadron operating against it for four different missions.[42]

Several years later, another lecture postulated:

> The results that are achieved by attack of the national economic structure are also cumulative

and lasting. They build up from day to day and from week to week so that the pressure that formerly has been imposed by military action over long periods of time, may, by this method, be concentrated into a short period, and still produce that intense suffering upon the civil populace that has been essential for the collapse of the national morale and the national will to continue with the war.[43]

ACTS clearly saw bombardment as a linear process. Although there was some allowance for repairs, the lectures emphasized the cumulative nature of the destruction caused by bombardment. In other words, superposition applied to aerial attacks.

In the end, the ACTS theories were the first to incorporate some concepts from the complex paradigm. More precisely, the industrial web theory was based on the notion of economies composed of many interlaced, interacting parts. Economies exhibited hierarchies at local, national, and international levels. However, the doctrine did not extend much further into the complex paradigm. Furthermore, it contained principles from the Newtonian paradigm, most notably the search for efficiency and the presumption that bombing effects would accumulate. The result was a doctrine that had one foot in the complex paradigm and the other in the Newtonian paradigm.

The industrial web doctrine carried over into the air plans and operations of WWII. In 1940, General Henry H. Arnold established the Strategic Air Intelligence Section, A-2, to provide intelligence data and analyses for the Air Corps. Major Haywood S. Hansell, Jr., led the section's strategic air intelligence and analysis effort. A former student and instructor at ACTS, Hansell applied the industrial web doctrine to economic-industrial-social analyses of the Axis powers. Not surprisingly, the analyses closely paralleled ACTS lectures. With regard to the German economy, the section focused on electrical power, steel, petroleum products, the aircraft industry, transportation, nonferrous metal supplies, machine tools, and food processing and distribution. Their studies led to target folders with aimpoints and bomb sizes.[44]

In July 1941, General Arnold created the Air War Plans Division. The division consisted of Colonel Harold L. George, Lieutenant Colonel Kenneth N. Walker, Major Laurence S. Kuter, and Major Hansell, all of whom had previously taught at ACTS. They were tasked with determining the forces and munitions required by the Air Corps for a potential war with the Axis. The division made the key assumption that the German economy was drawn taut and under extreme stress due to the war. The division produced its first plan, *AWPD/1: Munitions Requirements of the Army Air Forces to Defeat Our Potential Enemies*, in August 1941. The target sets in priority order were (1) the German Air Force, including aircraft factories, aluminum plants, magnesium plants, and engine factories; (2) electric power, including power plants and switching stations; (3) transportation, with a focus on rail and water; (4) petroleum, including refineries and synthetic plants; and (5) the morale of the German people. The web linkages were important to the planners:

> Many factors formed vital links in Germany's
> industrial and military might. The overriding
> question was, which were the most vital
> links? And among these, which were the
> most vulnerable to air attack? And from
> among that category, which would be most
> difficult to replace, or to "harden" by dispersal
> or by going underground? Each link in the
> chain had its own interconnecting links and
> the search had to be for the one or more
> keys to the entire structure.[45]

The "*one or more keys* to the *entire* structure" were, of course, bottlenecks. It is important to note the implied search for the most *efficient* bottlenecks in Hansell's comment. From the target sets, to the concern about the interconnectivities, to the search for bottlenecks, the influence of ACTS doctrine upon AWPD-1 is clear.

A year later, President Roosevelt asked the military for another estimate of future military requirements. Hansell directed the air estimate. The result was *AWPD/42: Requirements for Air Ascendancy, 1942.* There was no change in strategic objective, and the approach followed closely that used for AWPD-1. The list of targets changed slightly between AWPD-1 and AWPD-42, taking into account the current strategic situation and lessons learned from combat. Like its predecessor, AWPD-42 was accepted as strategic guidance even though it was a munitions plan. And once again, the heritage of ACTS doctrine was evident.

Following the Casablanca Conference in January 1943, a directive for a new combined strategic air plan was issued. The new air plan drew heavily upon the analyses of the Committee of Operations Analysts (COA), established two months earlier. The COA was composed of civilian industrialists and economists who examined in detail enemy economic and industrial structures. The committee used several guidelines to determine target priorities:

> In the determination of target priorities, there should be considered (a) the indis-pensability of the product to the enemy war economy; (b) the enemy position as to current production, capacity for production and stocks on hand; (c) the enemy requirements for the product for various degrees of activity; (d) the possibilities of substitution for the product; (e) the number, distribution and vulnerability of vital installations; (f) the recuperative possibilities of the industry; (g) the time lag between the destruction of installations and the desired effect upon the enemy war effort.[46]

The Committee looked for weaknesses and bottle-necks in the Axis economies that could be exploited. The analyses included the effects of substitution, in-dustrial recuperation, and time criticality of the product on the war effort. Their prioritized target lists bore close resemblance to both AWPD-1 and AWPD-42. Not surprisingly, the ensuing strategic air plan for the Combined Bomber Offensive was similar to its two predecessors.

Due to the influence of the industrial web doctrine, the air campaign plans of World War II contained elements from the complex paradigm. The planners considered economies to be composed of many linked, interdependent sectors. As Hansell noted, the Air War Plans Division was concerned with linkages and their effects. Linkages figured in COA analyses as well. The Enemy Objectives Unit (EOU), a separate advisory panel on target selection and intelligence in the 8th Air Force, examined the nature of linkages in some detail. In particular, the EOU developed a measure of the "depth" of a good or service.[47] Depth was roughly equivalent to the amount of time that would elapse between the cessation of production of some item and the appearance of its shortage on the battlefield. It is analogous to the notion of time criticality in tight and weak couplings.[48] The analysts applied this measure to items with direct military utility such as petroleum as well as items with indirect utility such as electric power. The various planning groups recognized the existence of higher order linkages and hierarchies. All of these concepts played into planning considerations.

Nevertheless, aspects of the planning were Newtonian. Two particular examples stand out. First, to a certain degree, targeting was reductionist. The planners selected target systems that were considered vital to the Axis war effort, such as electricity or petroleum. The linkages between these systems and other sectors of the economic-industrial-social fabric of the nation were important. Once the target systems were selected, the analysts often proceeded in a reductionist fashion: they considered the systems in isolation from one another, and selected aimpoints within the isolated systems. Thus, the targeteers generated "efficient" aim points for individual systems without detailed consideration of cross-system effects. A close examination of AWPD-1 and AWPD-42 shows little consideration of the effects of aimpoints in one target system on the operation of another. In reality, economies are highly interconnected with multiple, important cross-system effects that must be taken into account. A Newtonian, reductionist methodology will miss these effects; consequently, what appear to be efficient aimpoints for an isolated target set might be neither efficient nor effective when the economy as a whole is considered.

A second example of Newtonian thought is found in the planners' treatment of substitution. The EOU considered two types of substitution in their analyses.[49] The first entailed replacing processes and equipment, such as using kilns from the ceramics industry in place of destroyed kilns used to dry grinding wheels. The second focused on replacing products or services, such as using aluminum in electrical lines rather than copper so that the copper could be used in radios and communication equipment. Hansell noted the importance of substitution and workarounds but conceded that they were difficult to anticipate and analyze. In general, the planners focused on bottlenecks for which substitution was very difficult if not impossible.

However, the planners missed a crucial aspect of substitution. They focused on what Olson terms the "tactical supply problem" rather than the more important "strategic supply problem."[50] The tactical supply problem deals with depriving a military of specific items without which it cannot function. For example, if an airplane requires a certain part to operate properly (such as a variable pitch propeller spring), a lack of that part will ground it. No amount of other types of spares can make up for the missing part. Fundamentally, the tactical supply problem notes that a military cannot operate normally without an adequate supply of the right types of equipment and spares. The strategic supply question focuses on the ability of a nation as a whole to provide for the requirements of its military. In this case, Olson maintains that a nation can generally make up most of any shortages of required goods and services if it is willing and able to shift production from other items into production of the missing items. In essence, the strategic substitution question examines the adaptations an economy makes to fulfill the requirements of war.

Olson points out that the planners understood the tactical substitution problem, but failed to comprehend the implications of strategic substitution. By focusing on bottlenecks that would lead to tactical supply problems, the planners overlooked the ability of the German economy to perform strategic shifts. Consequently, they underestimated the ability of the Germany economy to absorb and adapt to the destruction wrought by the strategic air campaign. In a close analogy to complexity science, Olson states:

...a modern economy is not after all like a finely jeweled watch: it can lose a single part and continue to function. The modern economy is admittedly 'intricate and interdependent,' like an expensive watch, but the analogy stops there. For while the watch cannot replace one of its components, an economy can...in a modern economy change is a law of life, and adjustment to change a commonplace, in peace and in war. It would be much better to compare an economy with a tree, which can grow a new branch when an old one is removed, than to a building, which will collapse if part of its foundation is destroyed.[51]

The air planners of World War II viewed the Axis economies and the substitution problem from the Newtonian perspective rather than the appropriate complex viewpoint. They thereby locked an important aspect of the air campaign planning into the Newtonian paradigm.

Airpower theory and application was clearly shifting from the Newtonian to the complex paradigm by World War II. The industrial web theory incorporated some characteristics of complex adaptive systems. The ACTS faculty recognized that economies were composed of linked and interdependent sectors with different hierarchical levels. The lectures had hints of nonlinearities and feedback loops, and some detail on the nature of linkages. However, Newtonian characteristics abounded: important behaviors such as adaptation and their implications for warfare were not fully understood, targeting was largely reductionist, and planners focused on efficiency. The result was an airpower theory that was part Newtonian and part complex. Decades would pass before airmen would propose theories that more fully rested upon the complex paradigm.

Modern Airpower Thought

Several current airpower concepts have significantly influenced modern operational and strategic art. We will examine three of these theories, specifically the Five Rings model of Colonel John A. Warden III, the OODA loop developed by Colonel John R. Boyd, and parallel warfare. These concepts largely rely upon ideas intrinsic to complexity. In fact, Boyd is among the first military theorists to directly state that his concepts represent far-from-equilibrium, complex adaptive systems.

Colonel John A. Warden III's Five Rings model provides a conceptual framework for analyzing the physical aspects of an opponent.[52] Warden makes a clear distinction between the physical and morale sides of an enemy; he completely decouples one from the other:

> The advent of airpower and accurate weapons has made it possible to destroy the physical side of the enemy. This is not to say that morale, friction, and fog have all disappeared. It is to say, however, that we can now put them in a distinct category, separate from the physical. As a

consequence, we can think broadly about war in the form of an equation:

(Physical) x (Morale) = Outcome

...Looking at this equation, we are struck by the fact that the physical side of the enemy is, in theory, perfectly knowable and predictable. Conversely, the morale side— the human side—is beyond the realm of the predictable in a particular situation because humans are so different from each other. Our war efforts, therefore, should be directed primarily at the physical side.[53]

The decoupling of the physical and morale aspects of the enemy is a critical assumption in the theory. Warden then proceeds to examine only the physical characteristics and means of the enemy.

Figure 1 is a schematic of the Five Rings model. As the name implies, the model portrays an enemy system as a set of five concentric rings. The innermost ring represents the central leadership or direction of the system. The ring contains not only the enemy leaders, but also command communications and leadership security. These are the most crucial functions for normal operation of the enemy. The next most critical ring, organic essentials, contains key production (military and nonmilitary) and energy sources. The third ring encompasses the infrastructure of the system—its transportation networks, factories, and so forth. Generally, elements of this ring are more redundant and numerous than those of the innermost two rings. The fourth ring contains the population and the food sources of the enemy state. The number of

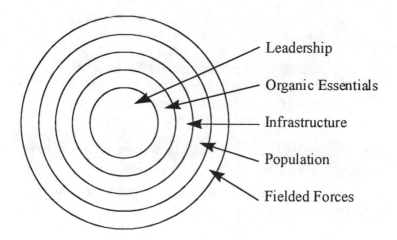

Figure 1. The Five Rings Model

targets in this ring are more numerous than those of the inner rings and often more difficult to attack, moral considerations notwithstanding. The outermost ring represents the fighting mechanism or fielded enemy forces. Targets in this ring are numerous and hard, compared to those in the other rings. Ideally, a strategic attack will focus on centers of gravity closest to the leadership or innermost rings. Warden maintains that the five rings model can be used to represent any type of organization that can operate autonomously, whether it be a human body, a drug cartel, an electrical grid, a guerrilla organization, or an enemy nation.

Warden ascribes a number of physical attributes to the enemy system that are characteristic of complex adaptive systems. First, the individual components—agents—of the rings are intertwined in complicated manners. To give a few examples, electrical power grids (ring two) feed communications networks (ring one), transportation networks (ring 3) provide coal for electrical generators (ring 2), and petroleum distribution networks (ring 2) deliver fuel oil to heat the homes of the population (ring 4). In general, the interactions are nonlinear. They may be of first or higher order. The nonlinear linkages lead to dynamical, often unpredictable behaviors when the enemy system is attacked, such as cascading breakdowns or possibly chaos. In sum, the enemy system consists of many agents linked together in intricate, nonlinear manners.

The second important physical characteristic of the model is an intrinsic, hierarchical structure. Warden contends that each element of every ring can itself be decomposed into five subrings. For example, an electrical grid (ring two) is composed of a leadership function (central control), organic essentials (energy input, such as coal, oil, or hydro power), an infrastructure (transmission lines and transformers), a population (workers), and a fighting mechanism (repairmen). Furthermore, each subring can be decomposed to an even finer degree. The result is a set of nested rings—a hierarchical organizational structure.

The third characteristic of Warden's model is the layered communications between and across hierarchical levels. Communications (linkages) may be horizontal between two agents in different rings, vertical between agents at different hierarchical levels, or some combination of the two. Like the industrial web, the myriad of communication paths creates an intricate fabric for information exchange, and opens the door to some complex behaviors.

Not only does the model have several of the physical characteristics of complex adaptive systems, it also may allow some of the nonlinear behaviors of such systems. It may be possible to drive the system to chaos or force a cascading breakdown of its normal operating state. In particular, attacks may render the enemy incapable of further resistance, a condition termed strategic paralysis. In this state, the enemy is no longer capable of adapting to changes in his environment. Metaphorically, the enemy system has been pushed away from the edge of chaos.

Despite the elements it shares with complex systems, the model has a definite Newtonian aspect. The key assumption that the physical and morale sides of warfare can be decoupled is reductionist and has significant implications. As Warden notes, it is in principle possible to know everything about the physical side of an enemy—the topology of his power grid, his communications systems, his transportation networks, and so forth, including interconnections. However, without the morale (human) side and the subsequent ability to adapt, the physical side reduces to a Newtonian machine much like Olson's finely jeweled watch. The human element in warfare is precisely what enables combatants to constantly adapt to their changing environments. It is indeed possible to deprive an enemy of his ability to react, as the Gulf War demonstrated. However, stripping the model of the morale side artificially removes much of the opponent's capacity to adapt from the onset. The assumption of separate morale and physical aspects has significant ramifications for the model, and gives it a decidedly Newtonian flavor.

In contrast to Warden, Colonel John R. Boyd's theories of warfare focus on processes and patterns of thought. He is perhaps best known for his Observe-Orient-Decide-Act (OODA) model of decision making.[54] In this model, a system observes some event of interest, decides how to resolve a problem posed by the event, and finally acts upon its decision. The cycle is illustrated in Figure 2. This process frequently arises during military operations, where the commander's objective is to "get inside the enemy's OODA loop." He does this by simultaneously destroying the enemy's capability to sense, process, and act on information while preserving his own ability to do so. Once reaching this point, the commander can force the enemy to constantly react rather than take the initiative. Both the friendly and enemy sides cycle

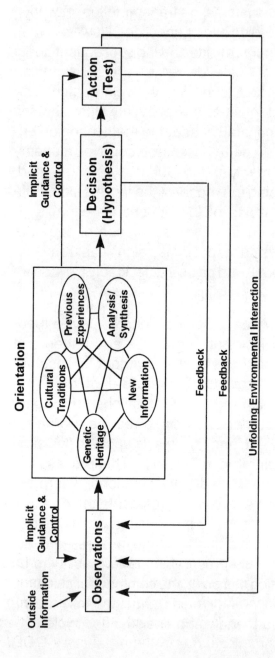

Figure 2. The OODA Loop

through the OODA loop; the friendly objective is to do so more rapidly than the adversary. In doing so, the enemy's actions lose coherence with the changing environment.

The OODA model has deep parallels to the manners by which complex adaptive systems process information and adapt to their environments. In an exceptionally close paraphrase of Boyd's theory, Kauffman describes how complex biological systems interact with their environments:

> But it is also plausible that systems poised at the boundary of chaos have the proper structure to interact with and internally represent other entities of their environment ...organisms sense, classify, and act upon their worlds. In a phrase, organisms have internal models of their worlds which compress information and allow action... Such action requires that the world be sufficiently stable that the organism is able to adapt to it. Were worlds chaotic on the time scale of practical action, organisms would be hard pressed to cope.[55]

By sensing, classifying, and acting upon their worlds, organisms cycle through Boyd's OODA loop and adapt to changing conditions. Chaos on the time scale of practical action implies that some organisms are unable to assimilate environmental conditions fast enough. Here, the environmental conditions change within a single period of an organism's OODA loop, precluding it from acting coherently. In essence, the environment has "gotten with the organism's OODA loop." The correspondence between Boyd's model and the adaptive and information processing capabilities of complex systems is striking.

Boyd recognizes that his model of warfare is rooted in the complex paradigm. He notes that the *"OODA Loop Sketch* and related *Insights* represent an evolving, open-ended, far-from-equilibrium process of *self-organization, emergence* and *natural selection."*[56] The OODA loop itself is a complex adaptive process, employed in conflicts by competing complex adaptive systems. As Boyd implies, the model lies outside the boundaries of the Newtonian paradigm, and falls within the complex paradigm.

Closely related to both Boyd's and Warden's models is parallel warfare. Warden asserts that states have on the order of several hundred vital, strategic level targets with perhaps ten aimpoints per target.[57] Furthermore, these targets are usually small, expensive, have few backups and are difficult to repair or replace. By attacking a large percentage of targets simultaneously, the damage to and effect upon the enemy can be overwhelming. The enemy will be incapable of reacting to and recovering from the damage. Strategic paralysis or collapse of the enemy may result from his inability to respond to a parallel attack. The opposite limit is serial warfare, in which only one or two targets are attacked at a time. Here, the opponent may have adequate time to recover from the attack, repair the damage, disperse his resources, and adapt. Parallel warfare is fundamentally different from serial warfare, in that force is no longer concentrated against one or two targets at a given moment of time. Rather, force is distributed widely throughout the enemy state, thus significantly complicating the problems for the defense and rendering an adequate response far more difficult.

Parallel warfare is metaphorically tied to adaptation of complex systems. The objective of parallel warfare is to so rapidly modify the environment that the enemy is incapable of reacting to the changes. Borrowing Boyd's terminology, the attack creates environmental change on a time scale shorter than the enemy's OODA cycle time. The opponent can no longer sense the changes, and consequently cannot react—adapt—appropriately. The attack has "gotten within the enemy's OODA loop."

Parallel warfare is crucial to the Five Rings model. To a certain degree, it enables Warden to decouple the physical from the morale side of warfare. By depriving the enemy of a coherent response, he can no longer repair or adapt to the destruction of his physical resources. In the theoretical limit of attacking *everything* simultaneously, the physical side of the enemy may indeed decouple from the morale and reduce to the finely jeweled watch. In the opposite theoretical limit of prolonged serial warfare, the physical and morale sides are inseparable. Warden's key assumption depends heavily upon the rapid environmental changes created by parallel warfare.

The modern theories of airpower we examined are more closely tied to complexity theory than their predecessors. The Five Rings model shares characteristics of complex adaptive systems, yet still has Newtonian aspects. When coupled with parallel warfare, it moves more closely to the complex paradigm. Boyd uses the language and concepts of complexity, recognizing that his models are indeed part of new paradigm. The impact of the new sciences is more clearly felt in modern airpower theories than in their earlier relatives.

Summary

Airpower theory is indeed transitioning from the Newtonian to the complex paradigm. The early theories of Douhet and Mitchell went beyond existing military thought, taking into consideration the revolutionary capabilities the airplane brought to warfare. Yet their concepts fell within the Newtonian paradigm. By the mid-1930s, the industrial web model incorporated some of the characteristics of complex adaptive systems. However, the behavioral aspects of complex systems were not clearly recognized, especially with respect to strategic substitution and adaptation. The theories of the WWII era were thus a mixture of complex and Newtonian ideas. In modern times, theoreticians have moved more closely toward the new sciences. Warden's Five Rings model coupled with parallel warfare incorporate many characteristics and behaviors of complex adaptive systems. Boyd's theories have made the transition, as he identifies his insights and the OODA loop with complex systems and their behaviors.

Application

The complex paradigm provides a powerful metaphor for airpower. We will only gain from its insights if we frame our theories within the new paradigm and apply those theories in combat. In the preceding section we examined existing theories for Newtonian or complex aspects, and categorized them accordingly. We will now take the next important step by applying concepts from complexity-based airpower theories to practice. As an example, we will examine targeting modern economic systems.

Economies are complex adaptive systems.[58] ACTS took the first steps toward this realization with its industrial web theory. However, we can go far beyond the web by examining the nature of its interconnections further. The agents in an economy are very intricately interwoven. The linkages have a variety of characteristics, including:

- nonlinearities;

- feedback and feedforward paths;

- serial, branching, and parallel processes;

- tight, moderate, and loose couplings; and

- a variety of interaction time scales.

The linkages lead to synergies and dependencies of various agents upon each other. Backup systems,

workarounds, repairs, tactical and strategic substitution, and human ingenuity provide the economy with
flexibility and adaptability in the face of disasters. An
economic system is without doubt complex and adaptive, and must be targeted accordingly.

Given this nature of an economy, what insight can the
complex paradigm provide the planner? Perhaps most
important is that a reductionist targeting approach is
limited and may overlook important emergent system
behaviors. We saw that targeting frequently followed
the reductionist methodology: the planner selects
target sets, then chooses aimpoints in each set without considering the linkages *between* sets. These
linkages are critical, as they give rise to cross-system
effects such as cascading breakdowns. By ignoring
the linkages, the planner suppresses these effects.
At best, the result is a campaign plan that inaccurately
portrays the system-wide effects of destroying the selected targets. At worst, the result is a plan that fails
to accomplish its objectives. What is required is a
holistic targeting methodology that incorporates the
linkages and synergies—an approach that targets the
economy as a whole.

Engineering analysis tools and computer simulation
techniques may provide the key to holistic targeting.
Industry regularly uses engineering analysis tools to
plan, design, and operate many sectors of an economy.
Electrical utilities use circuit analysis techniques to
manage their grids, develop contingency plans, and
design for future growth. Similarly, water utilities and
petroleum pipeline companies employ hydraulic

analyses in their daily operations. The same is true for other elements of an economy, including communications, transportation, and natural gas distribution, to name a few. Collectively, these engineering techniques are called nodal analyses. The important point is that nodal analyses can determine the effects of the loss of one or more elements in a network. By analogy, a military planner could use the same tools to determine the effects of destroying certain network elements. It may be possible to further extend this idea, by considering several sectors of an economy together and determining the cross-sector effects with coupled nodal analyses. Yet a further extension would couple complex adaptive computer optimization routines such as genetic algorithms to nodal analyses to "evolve" target sets inside a computer.[59] In this case, the planner would be targeting a complex adaptive economy with a complex adaptive system. By incorporating the linkages between sectors in an economy, targeting moves from reductionist to holistic, taking into deeper consideration the nature of the economy and the way it will respond to an attack.

In this example, we have gone beyond simple metaphorical comparisons of complexity science to airpower. Starting from the complex paradigm, we theorized that modern economies are complex adaptive systems. We then applied a principle from complexity science to a practical problem by examining holistic targeting. The result was a proposed methodology for targeting that may provide greater insight into the manner by which the enemy economy will respond to an attack. It is only in applying theory developed within the complex paradigm to practical problems that we can take advantage of what complexity science has to offer, and that we put the complex rubber on the ramp.

Summary

Warfare is indeed a nonlinear, complex, adaptive phenomenon with two or more coevolving competitors. Airpower theory has gradually moved to this viewpoint, albeit without conscious effort until recently. Early airpower theories were firmly locked in the Newtonian paradigm. The transition to the complex paradigm began at ACTS, with the development of the industrial web theory. The transition has accelerated in the last few years, to the point where complexity has been explicitly called upon in several airpower theories and analyses. The shift away from the Newtonian paradigm must continue, as the paradigm is too limited to capture the essence of warfare. Only through the complex paradigm will theorists and strategists fully understand and capitalize upon the complex, nonlinear nature of warfare.

We must, however, move beyond simple metaphorical comparisons or analyses within the complex paradigm. The insights from complexity will have their greatest utility in practical applications. We saw this in the example of targeting modern economies. To employ Newtonian techniques to an inherently complex phenomenon hampers flexibility, inhibits creativity, and significantly limits our comprehension of the operational environment. We must formulate airpower theories within the complex framework and use the resulting insights in operational applications. To remain locked in the Newtonian paradigm will only deny airpower its full potential in future conflicts.

End Notes

1. John F. Schmitt, "Chaos, Complexity & War: What the New Nonlinear Dynamical Sciences May Tell Us About Armed Conflict" (Quantico, VA: Marine Corps Combat Development Command, 4 September 95), 16-25.

2. E. C. Zeeman, "Catastrophe Theory," *Scientific American*, vol. 234, no. 4 (April 1976), 65-83; E. C. Zeeman, *Catastrophe Theory: Selected Papers, 1972-1977* (Reading, Mass: Addison-Wesley Publishing Company, 1977).

3. The chaos theory literature is vast. Excellent introductions include James Gleick, *Chaos: Making a New Science* (New York: Penguin Books, 1987); James P. Crutchfield *et al*, "Chaos," *Scientific American*, vol. 255, no. 6 (December 1986), 46-57; and Celso Grebogi *et al*, "Chaos, Strange Attractors, and Fractal Basin Boundaries in Nonlinear Dynamics," *Science*, vol. 238, no. 4827 (30 October 1987), 632-638.

4. A general introduction to complexity is M. Mitchell Waldrop, *Complexity: The Emerging Science at the Edge of Order and Chaos* (New York: Simon & Schuster, 1992). A more technical overview of the field is given in Grégoire Nicolis and Ilya Prigogine, *Exploring Complexity: An Introduction* (New York: W. H. Freeman and Company, 1989).

5. For applications of catastrophe theory to warfare, see Zeeman (1976), 76-77; Zeeman (1977), 16-17. Chaos and warfare are explored in Todor D. Tagarev et al., *Chaos in War: Is It Present and What Does It Mean?* (Maxwell AFB, AL: Air Command and Staff College, 1994); Pat A. Pentland, *Center of Gravity Analysis and Chaos Theory or How Societies Form, Function, and Fail* (Maxwell AFB, AL: Air War College, 1994). Applications of complexity theory to warfare are presented in Thomas J. Czerwinski, "Command and Control at the Crossroads," *Marine Corps Gazette* (October 1995), 13-15; Schmitt; and Steven M. Rinaldi, *Beyond the Industrial Web: Economic Synergies and Targeting Methodologies* (Maxwell AFB, AL: Air University Press, April 1995).

6. For example, we can expand the nonlinear system as a Taylor series, and keep only the constant and first-order (linear) terms. We then solve the linearized equations. This technique is frequently employed in the sciences and engineering.

7. Schmitt, 17-22.

8. The air campaigns in World War II generally followed this philosophy. See, for example, *AWPD/1: Munitions Requirements of the Army Air Forces to Defeat Our Potential Enemies*, 12 August 1941, USAF Historical Research Agency (hereafter HRA) file 145.82-1; and *AWPD/42: Requirements for Air Ascendancy, 1942*, HRA file 145.82-42.

9. Schmitt, 16-25.

10. Lee A. Segel, "Grappling with Complexity," *Complexity*, vol. 1, no. 2 (1995), 18-25.

11. The couplings between agents may be tight or loose, branching or sequential, and may contain feedback and/ or feedforward paths. Rinaldi, 8-9.

12. Roger Lewin, *Complexity: Life at the Edge of Chaos* (New York: Macmillan Publishing Company, 1992), 12-13, 47.

13. P. W. Anderson, "More is Different," *Science*, vol. 177, no. 4047 (4 August 1972), 393-6.

14. Nicolis and Prigogine, 13.

15. Stuart A. Kauffman, *The Origins of Order: Self-Organization and Selection in Evolution* (New York: Oxford University Press, 1993), 173.

16. Nicolis and Prigogine, Chapter 1.

17. Kauffman, 173; Lewin, 48-55; Nicolis and Prigogine, 8.

18. Kauffman, 173.

19. Compare to the Organization Process Model II in Graham T. Allison, *Essence of Decision: Explaining the Cuban Missile Crisis* (HarperCollins Publishers, 1971), Chapter 3.

20. Lewin, 15, 138; Waldrop, 145-47.

21. John R. Boyd, "A Discourse on Winning and Losing," unpublished briefing and essays, Air University Library, document no. MU 43947 (August 1987). In particular, see p. 5 of Chapter 1, "Patterns of Conflict."

22. Giulio Douhet, *The Command of the Air* (Washington, D.C.: Office of Air Force History, new imprint, 1983), 142.

23. Ibid, 58.

24. Ibid, 25, 142.

25. Ibid, 28-32. Note that his arguments must be taken in the context of the geostrategic position of Italy.

26. Ibid, 99, emphasis in original.

27. William Mitchell, *Skyways: A Book on Modern Aeronautics* (Philadelphia: J. B. Lippincott Company, 1930), 255.

28. Ibid, 253.

29. William Mitchell, *Winged Defense* (New York: Dover Publications, Inc., 1988, reprint of 1928 original publication), 10, 126-7.

30. William Mitchell, *Our Air Force: The Keystone of National Defense* (New York: E. P. Dutton & Company, 1921), xix, xxi.

31. Mitchell (1988 reprint), 133.

32. Douhet, 11-12.

33. Ibid, 17, 55, 218-219, 239. His argument, of course, does not foresee the development of radar and modern defensive air control concepts.

34. Ibid, 20-22, 35-41, 50.

35. One may argue that it is unfair to mention that Douhet's and Mitchell's theories did not include any concepts from complexity, as the science would not arise for another 60 years. However, as Beyerchen has eloquently argued, Clausewitz's theory of war included many concepts from nonlinear dynamics and chaos theory—two sciences developed some 130 years after his death. Alan Beyerchen, "Clausewitz, Nonlinearity, and the Unpredictability of War," *International Security*, vol. 17, no. 3 (Winter 1992/93), 59-90.

36. Major Muir S. Fairchild, "National Economic Structures" (Maxwell Field, AL: Air Corps Tactical School Lecture, 5 April 1939), 8-9.

37. "Principles of War Applied to Air Force Action" (Maxwell Field, AL: Air Corps Tactical School Lecture, 1934-1935), 3.

38. "Air Force Objectives" (Maxwell Field, AL: Air Corps Tactical School Lecture, 1934-1935), 5, 8.

39. Ibid, 2.

40. "Principles of War Applied to Air Force Action," 2-3.

41. Ibid, 5.

42. Ibid, 7-8. Emphasis in original.

43. Fairchild, 5.

44. Haywood S. Hansell, Jr., *The Air Plan that Defeated Hitler* (Atlanta: Higgins-McArthur/Longino & Porter, Inc., 1972), 49-51.

45. Ibid, 79.

46. Memorandum to Lieutenant General Arnold, 8 March 1943, Subject: Report of Committee of Operations Analysts with Respect to Economic Targets Within the Western Axis. Guido R. Perera, "History of the Organization and Operations of the Committee of Operations Analysts, 16 November 1942–10 October 1944," Vol. II, Tab 22, HRA file 118.01.

47. Carl Kaysen, "Note on Some Historic Principles of Target Selection," U.S. Air Force Project RAND Research Memorandum RM-189 (15 July 1949).

48. See footnote 11.

49. Kayson, 5.

50. Mancur Olson, Jr., "The Economics of Target Selection for the Combined Bomber Offensive," *RUSI Journal*, vol. CVII (November 1962), 308-314.

51. Ibid, 312.

52. John A. Warden III, "The Enemy as a System," *Airpower Journal*, vol. IX, no. 1 (Spring 1995), 40-55; John A. Warden III, "Employing Air Power in the Twenty-first Century," in *The Future of Air Power in the Aftermath of the Gulf*

War, edited by Richard H. Shultz, Jr. and Robert L. Pfaltzgraff, Jr. (Maxwell Air Force Base: Air University Press, July 1992), 57-82.

53. Warden (1995), 43.

54. Boyd, 5. He introduces his concept on page 5 of Chapter 1, "Patterns of Conflict," and develops it with historical examples.

55. Kauffman, 232. Although Kauffman is discussing biological systems, complex systems in other domains process information in essentially the same manner.

56. John R. Boyd, "The Essence of Winning and Losing," unpublished notes (28 June 1995), 4, emphasis in original.

57. Warden (1995), 54.

58. For a detailed examination of economies as complex adaptive systems and the implications for targeting, see Rinaldi.

59. Ibid, Chapter 4.

Chaos Theory and U.S. Military Strategy: A 'Leapfrog' Strategy for U.S. Defense Policy

Michael J. Mazarr

Applying chaos theory to U.S. military strategy and force structure is a perilous business. Some would doubt whether the theory has much meaningful application in social science at all. What, after all, are its recommendations? That rapid and discontinuous change is inevitable, the product of "sensitive dependence on initial conditions"? That we must be prepared for surprises? That we must be agile and flexible and quick on our feet?

If chaos theory is not to degenerate into an annoying repetition of the same themes, its practitioners must begin offering its practical lessons in a manner that can be understood by military planners. And its lesson is not, I should make clear, that the U.S. military needs to be ready for peacekeeping and other operations other than war in a "chaotic" post-cold war world; such short-term political chaos has very little to do with the vastly more profound and fundamental insights of chaos theory. No, if the theory is to make a real contribution to defense policy, it must do something more: without being determinative, it must point us in the direction of a coherent planning framework for U.S. military forces. I believe that it can do so, and in this paper I will explain how.

At the same time, at this point in its emerging application to the social sciences, chaos or complexity theory certainly cannot provide comprehensive answers. As Dr. Murray Gell-Mann stressed on the conference's first day, chaos theory remains in its formative stages; it is useful mostly as a spur to reconsider old ways of doing business and take seriously rapid and unpredictable change. My recommendations for force structure, for example, stem as much from an appreciation of accelerating change as from "complexity"—but chaos theory can help advocates of change make their case.

The Knowledge Era and International Relations

To begin with, it is noteworthy that the social and economic context of the post-cold war world parallels in important ways the kind of world described by chaos theory. In large measure this has to do with the emergence of a knowledge-based society, a transformation of social and economic life that is overturning the institutions and patterns and assumptions of the industrial era and substituting those of a new age.

There is a vast literature on the information or knowledge era, and I will not attempt to summarize its conclusions in any detail. Professor James Rosenau said a few words about this kind of world on the first day of the conference, and there are few better introductions to its character and implications than his own *Turbulence in World Politics*.[1] In brief, it involves the establishment of information and knowledge—their production, dissemination, storage, and use—as the fundamental social and economic activity, rather than the cultivation of agriculture or the production of manufactured goods. Perhaps the most powerful single measurement of the information sector's dominance is that service industries now represent something like 70 percent of the U.S. economy, both as a percentage of GNP and in terms of employment; manufacturing has declined to just over 20 percent. Not all services are knowledge-based, of course—but then, some manufacturing industries (computers, televisions) are tied to the knowledge sector. Estimates of the knowledge sector's component of the U.S. economy run in excess of 60 percent.

The knowledge era has a number of key hallmarks. As we have seen, it favors the transition from industrial manufacturing economies to service ones. In corporate organization, it allows and encourages decentralization, task and product teams, and ultimately new levels of "virtuality"; in management theory it points toward empowerment of workers and, again, democratization of decision making. It is global and local in scope at the same time—global in its reach, local in its focus, a paradox symbolized by multinational corporations with activities all over the world who nonetheless tailor their products to niche markets within individual countries. It is a world in which finance becomes more powerful than ever, challenging national central banks and international multilateral development banks for influence. It is an era in which old authorities are challenged and decay, and new or changed ones arise to take their place.

The knowledge era is therefore a time of rapid change, when old ways of doing business and the institutions that did that business fall to the side, in which new innovations can cascade very rapidly throughout an economy and society and create transformative change almost overnight. It is a time of rapid and discontinuous change, of small initial actions or innovations having dramatic and unforeseen implications. It is a time, in other words, in which chaotic models of social evolution come to the fore.

Responding to Chaos: Business Strategy

A number of thoughtful management experts have recent turned their attention to the implications of this new era for business. As one of the few avenues of productive strategic thinking in a chaotic mode, their advice is directly relevant to military planners trying to come to grips with the same currents of social change. Two writers in particular have done an especially good job of showing what the knowledge era, and its accompanying chaotic effects, mean for strategy: Richard D'Aveni and Gary Hamel. It seems to me that their ideas, while not explicitly intended for such a purpose, serve as a useful summary of the kinds of strategies required in a complex era.

This new era in business, much like the new era in international relations, is not simply one in which competition gives way to cooperation. These new forms of economic activity will hardly put an end to business competition. Indeed, they may be in the process of creating an unprecedented era of "hypercompetition," a phenomenon that mirrors many elements of complex systems and is examined in depth by Dartmouth Business School Professor Richard D'Aveni.[2]

Hypercompetition, D'Aveni contends, is "a condition of rapidly escalating competition based on price-quality positioning, competition to create new know-how and establish first-mover advantage, competition to protect or invade established product or geographic markets." The "frequency, boldness, and aggressiveness of dynamic movement by the players accelerates to create a condition of constant disequilibrium and change." D'Aveni's model is on display in the computer software industry, whose basic mode of operations has become a series of rapid competitive moves and countermoves that seek to create a series of *temporary* advantages. "Product life cycles and design cycles have been compressed," he writes, "and the pace of technological innovation has increased." So "instead of seeking sustainable advantage, strategy...now focuses on developing a series of temporary advantages. Instead of trying to create stability and equilibrium, the goal of strategy is to disrupt the status quo."

Later D'Aveni contends that "disrupting the status quo" should be the top corporate goal. In a hypercompetitive world, he writes, there will only be two kinds of companies: "the disruptive and the dead." D'Aveni's insightful approach has a number of powerful implications:

• Leapfrog or transformative strategies become more important than ever.

• Businesses will achieve smaller profit margins under the pressure of price wars.

• Trust will come under new pressure— and "once trust is lost, it's very hard to recapture, especially in global markets where xenophobia makes foreign competitors suspect."

• A "logical approach is to be unpredictable and irrational," so as to throw a competitor off their rhythm and distract them from your real intentions.

• Using the old strategy of attacking competitors' weaknesses "can be a
mistake"—because those weaknesses
won't last long, and you're shooting at a
moving target.

Another recent model of business strategy—London
Business School professor Gary Hamel's notion of
"strategy as revolution"—makes a very similar case.
Hamel argues that true business strategy "*is* revolution; everything else is tactics." Many firms, he argues,
"are reaching the limits of incrementalism"; pursuing
"incremental improvements while rivals reinvent the
industry is like fiddling while Rome burns." Companies like IKEA, the Body Shop, Dell Computer, and
Swatch are "shackled neither by convention nor by
respect for precedent" and are "intent on overturning
the industrial order." Never before, Hamel writes, "has
the world been more hospitable to industry revolutionaries and more hostile to industry incumbents. The
fortifications that protected the industrial oligarchy are
crumbling under the weight of deregulation, technological upheaval, globalization, and social change."[3]

One implication is that ideas that seem unusual should
get perhaps the best hearing of all. "Senior managers should be less worried about getting off-the-wall
suggestions," Hamel advises, "and more concerned
about failing to unearth the ideas that will allow their
company to escape the curse of incrementalism."[4] Another lesson of Hamel's perspective: rigid dividing lines
between industries are rapidly becoming obsolete. "Industry revolutionaries don't ask what industry they are

in. They know that an industry's boundaries today are about as meaningful as borders in the Balkans."[5] Finally, Hamel's principles of strategy suggest the need to empower workers. "Strategy making must be democratic," he writes, in part because the "capacity to think creatively about strategy is distributed widely in an enterprise. It is impossible to predict exactly where a revolutionary idea is forming; thus the net must be cast wide." Hamel refers to the need to "supplement the hierarchy of experience with a hierarchy of imagination."[6]

In sum, then, what advice does this new line of business thinking have for other social institutions in a complex, chaotic, fast-moving era? Strategies of the future will seek to disrupt the status quo and thrive in the resulting chaos. They will emphasize unpredictable moves. Incrementalism is a recipe for disaster. Authority must be decentralized and won by imagination and skill rather than seniority. Boundaries between disciplines will collapse. Managers must value new, unusual, what seem at first glance to be irrational suggestions.

Military Strategy: The Need for Revolutionary Thinking

To get a sense of how far the U.S. military is from a truly revolutionary response to the knowledge and information era, one need only hold D'Aveni and Hamel's advice up against the reality of military planning as we do it today. However much fast-paced, over-the-horizon, anti-traditional thinking—the kind demanded by the knowledge era—is going on in the military, that sort of mindset is clearly not guiding U.S. force structure planning today. In our quaint notion of a "hedge" against a Soviet Union that does not exist and our unreal (though undeniably comfortable) planning guide of "two (nearly) simultaneous regional contingencies," we are about as far away from out-of-the-box thinking as could be imagined.

Take, for example, our current approach to the Revolution in Military Affairs (RMA). In its true form, this concept represents the introduction of knowledge-era concepts and structures into warfare. And yet the existing DoD plan, at least in the medium-term, is not to achieve an RMA at all, but to graft elements of that revolution onto a military force still representative of industrial-era, attrition-style warfare.

Examples of this practice are easy to come by. A modern tank equipped with the global positioning system (GPS) and advanced cellular communications systems is not revolutionary, any more than an unstealthy attack aircraft with laser-guided bombs. A stealthy bomber raining cluster bombs on an advancing tank division is not revolutionary. Nor is an aircraft carrier equipped with fancy electronic countermeasures and radar detection systems. All of these capabilities—the capabilities on display in the Gulf War—represent evolutionary advances within the same mode of fighting that has prevailed, in some senses, since about 1940, and in others for hundreds of years.

Part of the confusion arises from the use of the term "information warfare," the term of art that attempts to capture the knowledge era's influence on war. Mastery and use of information is indeed at the core of the RMA. But this mastery does not simply involve adding one last bit of detail into a World War II-style tank outfit—as if, had Patton's tanks been equipped with the GPS, his divisions would have embodied the RMA. Rather, the *true* RMA represents an entirely new manner of warfare, using information, long-range precision strike, and other tools to destroy an enemy's ability and will to fight *without* closing on the battlefield and exchanging tank fire; *without* sending vulnerable aircraft deep into hostile airspace; and *without* deploying aircraft carriers close to an enemy coast.

The incrementalist notion of the RMA is ultimately self-defeating. It violates Gell-Mann's injunction that a period of rapid change is the time when it is most important to think comprehensively rather than narrowly. It indefinitely postpones the day when the U.S. military will truly depart from deeply-entrenched doctrines and routines and embrace the truly revolutionary elements of the new era in warfare. It guarantees that the lion's share of procurement and research and development funds will be devoted to slightly modified versions of weapons in regular use for almost a century. Incrementalism—the time-honored planning approach of every major bureaucracy everywhere—constitutes a mortal threat to our achievement of a true revolution in military affairs: If it is pursued bit by bit, added on to existing pre-RMA systems in appliqué fashion,[7] it will not be revolutionary at all. It will instead perpetuate old ways of conducting warfare and delay the time when the U.S. military enjoys the full advantages of the RMA.

The potential for a mortal threat to the RMA exists in part because of our budgetary predicament. As most U.S. military planners are now well aware, a crisis of defense policy is upon us, a crisis stemming from a simple, but lethal, mismatch between budgets, force structure, and modernization. The United States today has a small and steadily shrinking defense budget supporting a large force designed to fight two simultaneous regional wars. As a result, only modest amounts of long-term research and development or modernization are taking place. Not only does this situation make it impossible for the United States to implement the RMA in the coming decades; it makes it unlikely that we will maintain a high-quality, modern military force of any sort.

The numbers alone are startling—and for the most part, they are undisputed. No one denies the reality of the budget shortfall.[8] The force outlined in the Bottom-Up Review of Defense Priorities is underfunded by between $50 billion and $300 billion over five years. Put another way, to fully fund the BUR force, the United States ought to be spending in the neighborhood of 4 percent of GNP, while currently planned budgets will fall below 3 percent. This shortfall could manifest itself in three places: in force structure; in readiness; or in modernization. Because of the Clinton administration's military strategy of twin regional contingencies, it has felt unable to reduce force structure much beyond that of the Bush years. And because of the political and military costs of allowing combat readiness or training to slip, the administration has refused cutbacks in those areas as well.

As is now well-known, the result of these decisions has been to focus the effect of the budget shortfall on the third area of military spending: the United States has gutted modernization and research and development to pay for a relatively large, very ready force-in-being. Acquisition spending is down by 60 percent between 1987 and 1995. Research and development budgets will fall 40 percent from 1987 through 1999, and what is left focuses mainly on modifications and upgrades of existing systems rather than on developing new ones. The obvious consequence of slowed modernization is a military with aging equipment. By the year 2010, the average age of tanks in the U.S. military will be 21 years; of utility helicopters, nearly 30 years; of navy fighter aircraft, 15 years; of attack submarines and surface ships, 16 years; of air force fighter-attack aircraft, 20 years; and of air force bombers and transport planes, 35 years.[9]

These statistics tell a simple tale: the United States government has decided to mortgage the future of the military to its present. Slashing modernization in favor of force structure and readiness means a stronger military today in exchange for a weaker military tomorrow. "Modernization," General John Shalikashvili has said, "is tomorrow's readiness"[10]—and it is the

only route to the RMA. Without R&D and procurement, without new investments in tomorrow's military
in addition to mortgage payments on today's, the RMA
will never become a reality.

This kind of strategy would make sense if the United
States faced immediate and serious threats that man-
dated a very large, very ready military. But this is not
the case; the United States does not now face a major global rival, and will not face one for at least several
years. Regional predators like Iran and North Korea
will succumb to a much smaller U.S. force, and the
threat they pose is blatant enough and far enough outside the mainstream of world politics that we can
expect to assemble coalition efforts to defeat these
aggressors. On the other hand, ten years from now
we might face much more serious military threats. The
predators, if they still exist in their present, hostile form,
may be stronger, with new weapons and larger militaries. And one or another major power may undertake
a path toward regional aggression. "Our most serious" threats, says Columbia professor Richard Betts,
"will come down the road rather than tomorrow morning."[11] There is much to be said for constraining
existing capabilities to invest in modernization that
would produce a stronger military ten years hence.

The Need for a Leapfrog Strategy

Such an approach is available through what this essay will term the "leapfrog strategy." Its core idea is
simple: the United States should free up additional
money for investments in future defense capabilities,

by reducing its force structure and continuing to budget the planned increase in modernization funds beginning in FY1997; and it should invest that money, as well as the lion's share of existing procurement budgets, in RMA technologies, skipping one generation of advanced weapons systems now slated for procurement. In the process it should take the advice of D'Aveni and Hamel and treat strategy and force structure as revolutionary notions; in the fast-moving knowledge era, standing still invites disaster. By abandoning the idea of incremental modernization and striking out toward a truly new generation of weapons, the leapfrog strategy forces U.S. defense planners to abandon their appliqué model of the RMA and rethink doctrine, organization, and strategy from the ground up.

Currently, the Defense Department intends to purchase weapons over the next ten years that represent largely evolutionary advances over existing systems. Thus DoD will spend, in 1996 and 1997 alone, a billion and a half dollars to upgrade the M1 tank and the Bradley fighting vehicle; nearly half a billion dollars on a new artillery piece and its supply vehicle; and $400 million on a light, direct-fire tank. It will spend $500 million on the Comanche helicopter; $2 billion on the V-22 tilt-rotor aircraft; billions of dollars on new aircraft carriers and frigates; over $2 billion on various new models of the F/A-18 fighter-bomber; and roughly another $2 billion on new or upgraded F-15 and F-16 aircraft and the roots of a new generation of tactical combat planes.[12]

Yet, in the context of the RMA, many of these systems are doomed to obsolescence. Stealthy aircraft are of course an element of the RMA. But large surface ships, heavy armored vehicles, and non-stealthy aircraft will in coming years simply serve as magnets for advanced precision-guided weapons—and, perhaps, weapons of mass destruction as well. The truly advanced warfare of the twenty-first century will not be fought by aircraft carriers, tanks, and fighters as we now understand them, but by a very different sort of military force based around the principles of the RMA—speed, agility, synergy, information dominance, and lethal, long-range precision strike.

In many ways, the traditional systems can be thought of as a provisional generation of military technology, trapped between the highest expressions of pre-RMA military systems and the RMA itself. They might be called the Neanderthal Generation because, in an evolutionary sense, they are akin to Neanderthal Man: highly advanced, extremely intelligent, but doomed to extinction as a truncated line on the evolutionary tree. Recognizing these facts, the leapfrog strategy would skip this generation of technology in favor of a research and development and procurement strategy designed to bring the Revolution in Military Affairs into being by the year 2010.

This is not to suggest that the Defense Department is ignoring *all* the technologies relevant to the RMA. Nor is the argument here that *none* of the systems planned for deployment in the decade represent the nature of the RMA—a number of advanced munitions and pilotless drone aircraft now under development are well within the emerging style of warfare. The argument is simply that the balances are out of proportion: too much money is being spent on force structure and readiness rather than modernization at a time of reduced danger; and too many of our limited procurement dollars are being invested in the systems that symbolize a declining era in warfare.

What are the declining systems? If it committed itself to a leapfrog strategy, the United States would decide today that it had built its last heavy main-battle tank. It would have purchased its last unstealthy fighter or bomber aircraft; and perhaps, if we are especially bold, its last manned combat aircraft as well. With the C-17, it would have designed its last large transport aircraft. And the vessels now in dry-dock would represent the last aircraft carriers and other large surface combat ships built for the U.S. military. All of these systems belong to the Neanderthal Generation of military technology.

What new weapons and combat systems would take their place? The full answer to this question will only emerge over time, as research and development proceeds, and as the result of a careful process of evaluation within the Department of Defense—or, perhaps better for bureaucratic reasons, by a blue-ribbon commission outside the Pentagon. Nonetheless, some obvious areas of emphasis in the RMA include the following: the full range of information warfare capabilities, including computer hacker operations; long-range, precision-guided munitions; stealthy

aircraft; stealthy naval vessels, including both submarines and small, cheap, low radar cross-section,
PGM-firing surface ships; all-weather sensors and
targeting systems; drone observation aircraft and robotic ground fighting vehicles; a whole range of
non-lethal weapons; and others.

The leapfrog strategy is therefore a simple idea drawn
from the unavoidable situation which U.S. defense
policy makers confront. Faced with a rapidly-chang-
ing world whose evolution is more and more
resembling elements of chaos theory, the enormous
potential advantages of the RMA, budgets insufficient
to pay for even current forces, and barren modernization plans devoted largely to an improvement over
pre-RMA ways of doing business—faced with this unprecedented conjunction of factors—the choice for the
United States is obvious. Scrap the Neanderthal Generation; reduce force structure by perhaps 25 percent
to free additional resources; and design an investment
strategy to bring the Revolution in Military Affairs into
being by the year 2010.

Questions and Risks

Obviously, any approach as brazen as this will have
its share of risks and uncertainties. Understanding
and appreciating those risks will be a critical element
to implementing the leapfrog strategy in a sound manner. It would be nice, of course, to do both, but the
realities of our budgetary predicament will not allow
us that luxury.

Initially, U.S. military planners will need to inventory the capabilities they will be surrendering by scrapping the Neanderthal Generation of systems. If we stop building aircraft carriers today, for example, when would the U.S. carrier inventory drop to a level that would make it unable to maintain forward presence coverage in key regions of the world? If we cease buying tanks, roughly at what point would U.S. M-1s become unserviceable? The concept at issue is that of a *window of vulnerability*. Would the leapfrog strategy leave the United States with a decrepit force for two or three years, or five or ten, before the RMA systems actually came on line? Would our carriers become unusable before we possessed the intercontinental precision-strike capabilities to substitute for them?

Of course, the idea of a window of vulnerability is hardly unique to a leapfrog strategy. Current U.S. defense policy, and in particular its small procurement budgets, are already creating one. The only question is whether we address that risk by waiting until the last moment and then rushing a new set of Neanderthal Generation weapons into production, thus wasting resources and energy on a doomed class of combat systems; or whether we lay out a careful plan to close the window of vulnerability by realizing the RMA before it opens.

Moreover, the leapfrog strategy as I have outlined it does contain a substantial insurance policy against the transition. This insurance comes in the form of the modified Neanderthal Generation systems—

stealthy aircraft, stealthy robotic ships, unmanned air-craft—that were included in my list of RMA technologies. It is highly likely that, within our life-times, the process we now understand as the RMA will ultimately lead us to wars that have even out-stripped those space-age weapons. In purchasing them, however, we would preserve some degree of ability to fight "traditional" major wars, an ability reas-suring to U.S. friends and allies and cautioning to potential U.S. adversaries.

A second risk involves our level of certainty that we can bring operational RMA systems into the force in the next fifteen years. Is it possible to overcome the technological hurdles in these areas and produce sys-tems that work in that time frame? Or would we risk rushing into the force a series of ill-tested weapons prone to breakdown and failure? The state of tech-nology, and its rapid advance, suggest that the technological bridges can be crossed; none of the RMA systems contemplated here requires any pro-found new scientific breakthroughs. The marriage of an intercontinental-range missile and a precision-guided warhead is a matter of engineering rather than scientific research. Nonetheless, U.S. defense plan-ners must take careful stock of RMA systems and determine if they could be deployed in a sound man-ner by the year 2010.

Third, there is what might be called the "dreadnought fallacy": when a militarily dominant nation deploys a new generation of technology that renders previous ones obsolete, it can wipe away its advantage and begin a new arms race from scratch in an area where others can suddenly pull ahead. Some argue that this happened to Great Britain in the early twentieth century—the dreadnought trumped all previous fighting ships, and when others began building the huge new armored vessels, Britain's century-old dominance at sea came to a rapid end.

The obvious response to this argument is that no nation-state, not even one that is militarily dominant, can stop the progress of technology. If it chooses not to pursue a new generation of weapons, it will only be left further behind when others begin to exploit them. Had Britain chosen not to deploy dreadnoughts, other nations would without question have eventually deployed similar ships that would have rendered Britain's aging fleet useless. And the same is true today: the technologies that make up the RMA do exist; they will be developed, especially because so many of them overlap with emerging civilian applications; and nations will begin integrating these capabilities into their armed forces. The only question is whether the United States moves first to master them or is left behind.

Fourth and perhaps most fundamentally, leaders of U.S. defense and foreign policy must discuss the implications of an RMA force very carefully with our friends, allies, and potential adversaries. U.S. officials will need to reaffirm to all of them the effects of the RMA—most fundamentally, by reiterating that the purpose of the leapfrog strategy is to lay the foundation for another century of American leadership abroad; and to remind allies that, without such a strategy, the gradual decay of U.S. military capabilities is inevitable—just as, in the manner that business strategists understand, corporations that stand still in the global marketplace face inevitable decline.

Chaos Theory and a Stronger Military in 2010

The leapfrog strategy proposed here is not a radical, reckless approach to U.S. defense planning over the next ten to fifteen years. Given doctrinal barriers and budgetary shortfalls, a leapfrog strategy is simply the only way—short of a major upsurge in the defense budget—to make the RMA a reality in the foreseeable future. Circumstances and past defense decisions have left us with two clear, stark alternatives: leave the long-term defense program the way it is, and be witness to the steady erosion of U.S. military power; or adopt something like the leapfrog strategy and restore U.S. leadership for the better part of another century. If we take seriously the implications of chaos theory, there can be no other choice.

The leapfrog strategy is not without its risks and pitfalls. No approach to the large, complex issue of U.S. defense policy will be. But it is the one policy that recognizes the true value of the RMA and takes the steps necessary to bring it into being. As such, the leapfrog strategy is the single most fundamental organizational and strategic concept necessary to realize the RMA's full potential.

We have no time to waste. Every passing year exacerbates the deficit in defense modernization we are accumulating. If we are to avoid a serious window of vulnerability and bring the RMA into the force before our existing combat systems simply stop working, we must act rapidly and implement the leapfrog strategy before it is too late.

End Notes

1. James N. Rosenau, *Turbulence in World Politics* (Princeton, NJ: Princeton University Press, 1990).

2. These citations are drawn from Richard D'Aveni, *Hypercompetition: Managing the Dynamics of Strategic Maneuvering* (New York: The Free Press, 1994), pages xiii-xiv, 4-11, 18, 99-100, 225-28, 341-42, 349; his analysis of antitrust is on pp. 357-90.

3. Gary Hamel, "Strategy as Revolution," *Harvard Business Review*, Vol. 74, No. 4 (July-August 1996), pp. 69-71.

4. *Ibid.* p. 82.

5. *Ibid.*, p. 73.

6. *Ibid.*, pp. 75-76.

7. I am indebted to Dr. Dan Gouré of CSIS for this phrase.

8. These figures are drawn from Don M. Snider, Daniel Gouré, and Stephen Cambone, Project Directors, *Defense in the Late 1990s: Avoiding the Train Wreck* (Washington, D.C.: Center for Strategic and International Studies, 1995), pp. 9-15; and Dov Zakheim, "A Top-Down Plan for the Pentagon," *Orbis*, Vol. 39, No. 2 (Spring 1995), pp. 173-180.

9. The Defense Department has planned a 47-percent increase in procurement funds starting in FY1997, but this

plan assumes savings through base closures and management reforms that may never materialize. Nor will this money be enough to offset the substantial cuts already made. And I will argue below that the new funds would be better invested in a truly new generation of military capabilities than the upgrade-level technologies for which it is slated.

10. Cited in David C. Morrison, "Ready for What?" *National Journal*, Vol. 27, No. 20 (May 20, 1995), p. 1218.

11. Cited in *Ibid.*, p. 1219.

12. These figures are drawn from the relevant modernization sections of William J. Perry, *Annual Report to the President and Congress* (Washington, D.C.: U.S. Department of Defense, February 1995).

Contributors

ALAN D. BEYERCHEN is an associate professor of history at Ohio State University, and a fellow of the American Association for the Advancement of Science. A former Army captain, he is a specialist in German history and science. His *Scientists Under Hitler* received a Choice award as an Outstanding Book of 1977. In 1992, Dr. Beyerchen authored "Clausewitz, Nonlinearity, and the Uncertainty of War."

ZBIGNIEW BRZEZINSKI is Counselor for the Center for Strategic and International Studies and Professor of American Foreign Policy at the Johns Hopkins University School of Advanced International Studies. From 1977 to 1981, Dr. Brzezinski served as national security advisor to the President of the United States.

MURRAY GELL-MANN is a Nobel Laureate in Physics. He is a professor at the Santa Fe Institute and co-chairman of the Science Board. He is also affiliated with the Los Alamos National Laboratory and the University of New Mexico. He is the author of numerous publications, including the best-seller *The Quark and the Jaguar*, an introduction to Complexity for the general reader.

ROBERT JERVIS is an Adlai E. Stevenson professor of international affairs at Columbia University. Dr. Jervis is a fellow of the American Association for the

Advancement Of Science and of the American Academy of Arts and Sciences. In 1990 he received the Grawemeyer Award for his book *The Meaning of the Nuclear Revolution.* Dr. Jervis serves on the board of nine scholarly journals, and has authored over 70 publications.

STEVEN R. MANN is the U.S. Department of State's country director for India, Nepal, and Sri Lanka. He is a senior Foreign Service Officer with the rank of Counselor. Mr. Mann is a 1991 Distinguished Graduate of the National War College. He is the author of "Chaos Theory and Strategic Thought."

ROBERT R. MAXFIELD is a consulting professor at Stanford University, vice-chairman of the board of trustees of the Santa Fe Institute, a governor of Rice University, and serves on the board of a number of corporations. He has authored and co-authored several papers, including "Foresight, Complexity and Strategy."

MICHAEL J. MAZARR is editor of the *Washington Quarterly* and director of the New Millennium Project at the Center for Strategic and International Studies. He is also an adjunct professor at Georgetown University. A former Naval intelligence officer, Dr. Mazarr has authored several books, and is a regular columnist for the South Korean paper *Toyo Shinmum.*

STEVEN M. RINALDI is a Major and a developmental engineer at Headquarters Air Force Materiel Command. He is responsible for coordinating and planning cooperative science and technology programs with France, Russia, and Italy. While attending the School of Advanced Airpower Studies, Major Renaldi authored "Beyond the Industrial Web," a targeting approach employing Complexity theory.

JAMES N. ROSENAU is University Professor of International Affairs with the George Washington University. His scholarship has focused on the dynamics of change in world politics and the overlap of domestic and foreign affairs, resulting in more than 35 books and 150 articles, including *Turbulence in World Politics: A Theory of Change and Continuity*.

ALVIN M. SAPERSTEIN is a professor of physics and fellow of the Center for Peace and Conflict Studies at Wayne State University. He is a fellow of the American Physical Society and of the American Association for the Advancement of Science. Dr. Saperstein has published several books and articles on physics, energy, environment, military strategy and tactics, arms control, and dynamical modeling of international relations.

JOHN F. SCHMITT is a military consultant and writer. A major in the U.S. Marine Corps Reserve, he has been closely associated with Marine Corps doctrine since 1986. His credits include the manuals for *Ground Combat Operations*, *Warfighting*, *Campaigning*, *Command and Control*, and *Planning*, as well as the book *Mastering Tactics*. Major Schmitt continues to lecture at the National Defense University, the Marine Corps Schools in Quantico, and elsewhere.

EDITORS

DAVID S. ALBERTS is both deputy director of the Institute for National Strategic Studies, and director of Advanced Concepts, Technologies, and Information Strategies (ACTIS) at the National Defense University, which includes responsibility for the School of Information Warfare and Strategy and the Center for Advanced Concepts and Technology. He also serves as the executive agent for the Department of Defense's Command and Control Research Program.

THOMAS J. CZERWINSKI is a professor in the School of Information Warfare and Strategy, National Defense University. He has developed and teaches courses on complexity theory and its relationship to military strategy and operations, as well as publishing in the field.

Complexity And Chaos: A Working Bibliography

School of Information Warfare and Strategy

National Defense University

Washington, D.C.

BOOKS AND MONOGRAPHS

Anderson, Philip W., Kenneth J. Arrow, and David Pines, eds. *The Economy as an Evolving Complex System.* Redwood City, CA: Addison-Wesley, 1988.

Arthur, Brian W. *Increasing Returns and Path Dependence in the Economy.* Ann Arbor: University of Michigan Press, 1994.

Ashby, Ross W. *An Introduction to Cybernetics.* London: Chapman & Hall, 1965.

Athens, Arthur. *Unraveling the Mystery of Battlefield Coup d' Oeil.* Ft. Leavenworth, KS: School of Advanced Military Studies, Army Command and General Staff College, 1993.

Analyzes the current theories of intuitive decision-making from the fields of psychology, political science, cognitive science and management science, and concludes that the military should adopt an aggressive plan for educating its ranks in intuitive decision-making.

Atkins, P. W. *The 2nd Law: Energy, Chaos, and Form.* New York: Scientific American Books, 1994.

"All natural change is subject to one law. It's the second law of thermodynamics. In this volume, the acclaimed chemist and science writer P. W. Atkins shows how this single, simple principle of energy transformation accounts for all natural change...full of vivid examples, ideas, and images—but virtually no mathematics."

Axelrod, Robert. *The Evolution of Cooperation.* New York: Basic Books, 1984.

Barker, Patrick K. *Avoiding Technologically-Induced Delusions of Grandeur: Preparing the Air Force for an Information Warfare Environment.* Institute for National Security Studies, U.S. Air Force Academy, 1 October 1996.

Barnsley, Michael F. *Fractals Everywhere, 2nd Ed.* Cambridge, MA.: Academic Press Professional, 1993.

Bassford, Christopher. *Clausewitz In English: The Reception of Clausewitz in Britain and America 1815-1945.* New York: Oxford University Press, 1994.

See Chapter 2 for Clausewitz as a non-linearist.

Beaumont, Roger. *War, Chaos and History.* Westport, CN: Praeger, 1994.

Beer, Stafford. *Brain of the Firm: The Managerial Cybernetics of Organization.* Chichester, NY: J. Wiley, 1981.

Behe, Michael. *Darwin's Black Box: The Biochemical Challenge to Evolution.* New York: Free Press, 1996.

. "The result of biochemical investigation of cellular mechanisms, according to Behe, "is a loud, clear, piercing cry of 'Design!'" (See Johnson, Philip E; and Pearcey, Nancy; for book reviews.)

Bjorkman, Eileen, et al. "Chaos Primer," in *Air Campaign Course 1993: Research Projects,* edited by Richard Muller, Larry Weaver, and Albert Mitchum. Maxwell AFB, AL: Air Command and Staff College, 1993.

Bohm, David. *Wholeness and Implicate Order.* London: Routledge and Kegan Paul, 1980.

The universe must be fundamentally indivisible, a "flowing wholeness" in which the observer cannot be essentially separated from the observed.

Boyd, John R. "A Discourse on Winning and Losing," Unpublished briefing and essays, Air University Library, Document no. MU 43947, August 19897.

Boyd, John R. "Destruction and Creation." Unpublished paper. September 3, 1976.

Brillouin, Leon. *Science and Information Theory.* Academic Press, 1956.

Briggs, John and F. David Peat. *Turbulent Mirror.* New York: Harper & Row, 1989.

A very good introduction to nonlinear dynamics.

Campbell, D., R.E. Ecke and J.M. Hyman, eds. *Nonlinear Science: The Next Decade.* Cambridge, MA: MIT Press, 1992.

Capra, Fritjof. *The Turning Point: Science, Society, and the Rising Culture.* New York: Bantam Books, 1983.

"A compelling vision of a new reality. A reconciliation of science and the human spirit for a future that will work."

Casti, John L. *Complexification: Explaining a Paradoxical World Through the Science of Surprise.* New York: Harper Collins, 1994.

"Explores several types of phenomena, including the catastrophic, the chaotic, the paradoxical, the irreducible, and the emergent, and shows how these phenomena encompass...science, the arts, nature, the economy, and everyday life." See McKergow, Mark for book review.

Cohen, Jack and Ian Stewart. *The Collapse of Chaos.* New York: Penguin Books, 1994.

See Shepard, Harvey for book review.

Coveney, Peter and Roger Highfield. *Frontiers of Complexity: The Search for Order in a Chaotic World.* New York: Fawcett Columbine, 1995.

See Horgan, John. "A Theory of Almost Everything," for book review. Also, Lloyd, Seth. "Complexity Simplified."

Cramer, F. *Chaos and Order: The Complex Structure of Living Systems.* New York: VCH Publishers, 1993.

"...I do this in the hope that I might help build a bridge between science and technology on the one side and philosophy and art on the other."

Davies, Paul. C. W., ed. *The New Physics.* New York: Cambridge University Press, 1989.

Contains a number of survey articles on nonlinear dynamics and self-organization.

Davis, Lawrence, ed. *Genetic Algorithms and Simulated Annealing: An Overview.* Los Altos, CA: Morgan Kaufmann Publishers, Inc., 1987.

Dawkins, Richard. *Climbing Mount Improbable.* New York, NY: Norton, 1996.

Another defense of Darwinism. (See Johnson, Phillip E. for book review.)

De Landa, Manuel. *War in the Age of Intelligent Machines.* New York: Zone Books, 1991.

A quirky, but arresting, analysis of the relationship between Chaos theory, technology and warfare.

Department of the Navy, U.S. Marine Corps. *Warfighting,* FMFM-1. Washington, D.C., 14 June 1993.

Department of the Navy, U.S. Marine Corps. *Command and Control,* MCDP-6. Washington, D.C., 4 October, 1996.

Dockery, John T. and A.E.R. Woodcock, editors. *The Military Landscape: Mathematical Models of Combat.* Cambridge, U.K.: Woodhead, 1993.

   A mathematical treatise on the application of Catastrophe theory to combat command and control modeling.

Dorner, Dietrich. *The Logic of Failure: Why Things Go Wrong and What We Can Do to Make Them Right.* New York: Holt & Co., 1996.

   "People court failure in predictable ways, and failure does not strike like a bolt from the blue; it develops gradually according to its own logic." See Piattelli-Palmarini, Massimo for book review.

Gell-Mann, Murray. *The Quark and the Jaguar.* New York: W.H. Freeman, 1994.

Fischer, Michael E. *Mission-Type Orders in Joint Air Operations.* Maxwell AFB, AL: Air University Press, May 1995.

Gleick, James. *Chaos: Making a New Science.* New York: Viking, 1987.

   "A comprehensive introduction to the concepts of Chaos, yet popularized and written expressly for the nonmathematical laity."

Goldberg, David E. *Genetic Algorithms in Search, Optimization, and Machine Learning.* Reading, MA: Addison-Wesley Publishing Company, 1989.

Goodwin, Brian. *How the Leopard Changed its Spots.* London: Weidenfeld and Nicolson, 1994.

Gore, John. *Chaos, Complexity, and the Military.* Washington, DC: National War College, National Defense University, 1996.

"This paper briefly explains the key concepts behind chaos and complexity theory, looks at some of the efforts to apply them to military analysis, examines criticisms of these theories, and draws some implications from them for the military in the future."

Guastello, Stephen. *Chaos, Catastrophe, and Human Affairs: Application of Nonlinear Dynamics to Work, Organizations, and Social Evolution.* Mahwah, NJ: Lawrence Erlbaum Associates, 1995.

Hayek, F.A. *Law, Legislation and Liberty.* London: Routledge and Kegan Paul, 1982.

Hayek, F.A. *Individualism and Economic Order.* Chicago, IL: University of Chicago Press, 1948.

Holland, John M. *Hidden Order: How Adaptation Builds Complexity.* Reading, MA: Addison-Wesley Publishing Co., 1995.

See Horgan, John. "A Theory of Almost Everything, "Lloyd, Seth, "Complexity Simplified," and Carey, John "Can the Complexity Gurus Explain It All?" for book reviews.

Horgan, John. *The End of Science: Facing the Limits of Knowledge in the Twilight of the Scientific Age.* Reading MA: Addison-Wesley, 1996.

See Kelly, Kevin; Johnson, Philip; and Park, Robert L. for book reviews.

James, Glenn E. *Chaos Theory: The Essentials for Military Applications.* Newport Paper No. 10. Newport, R.I.: Naval War College, October 1996.

A veritable textbook written for "the broad population of students attending the various war colleges."

Johnson, George. *Fire in the Mind: Science, Faith and the Search for Order.* New York: Alfred A. Knopf, 1996.

See Lloyd, Seth, " Complexity Simplified" and Kepler, Tom for book reviews.

Kauffman, Stuart A. *The Origins of Order: Self-Organization and Selection in Evolution.* New York: Oxford University Press, 1995.

Kauffman, Stuart A. *At Home in the Universe: The Search for Laws of Self-Organization.* New York: Oxford University Press, 1993.

See Horgan, John. "A Theory of Almost Everything," Lloyd, Seth, "Complexity Simplified." and Carey, John "Can the Complexity Gurus Explain It All?" for book reviews.

Kellert, Stephen H. *In the Wake of Chaos: Unpredictable Order in Dynamical Systems.* Chicago: Chicago University Press, 1993.

Kelly, Kevin. *Out of Control: The Rise of Neo-Biological Civilization.* Reading, MA: Addison-Wesley, 1994.

See Taylor, William C., Bennahum, David S., and McKergow, Mark for book reviews.

Kelly III, Patrick. *"Modern Scientific Metaphors of Warfare: Updating the Doctrinal Paradigm."* Monograph, School of Advanced Military Studies, U.S. Army Command and General Staff College, Fort Leavenworth, KS, Second Term Academic Year 1992-93 (DTIC AD-A264 366).

Kiel, L. Douglas. *Managing Chaos and Complexity in Government.* San Francisco: Jossey-Bass Inc., 1994.

Koenig, John A. *A Commander's Telescope for the 21st Century: Command and Nonlinear Science in Future War.* Quantico, VA: Command and Staff College, Marine Corps University, 22 April 1996.

"While not directly related to each other, nonlinear theoretical understanding of the mind coincides with the advent of Naturalistic Decision Making. Together they provide useful insights into the basis of creativity and intuition in war that we can capitalize upon to educate future commanders."

Krasner, Saul, ed. *The Ubiquity of Chaos.* Washington, D.C.: American Association for the Advancement of Science, 1990.

Krugman, Paul. *Peddling Prosperity.* New York: W.W. Norton, 1994.

See Ch. 9: "The Economics of Qwerty," for thoughts on the economics of W. Brian Arthur.

Lam, Lui and Vladimir Naroditsky, eds. *Modeling Complex Phenomena.* New York: Springer-Verlag, 1992.

Langton, Christopher G., ed. *Artificial Life.* Santa Fe Institute Studies in the Sciences Of Complexity. Proceedings vol. 6. Redwood City, CA: Addison-Wesley, 1989.

Langton, Christopher G., Charles Taylor, J. Doyne Farmer, and Steen Rasmussen, eds. *Artificial Life II.* Santa Fe Institute Studies in the Sciences Of Complexity. Proceedings vol. 10. Redwood City, CA: Addison-Wesley, 1989.

Langton, Christopher G., ed. *Artificial Life III.* Santa Fe Institute Studies in the Sciences Of Complexity. Proceedings vol.17. Redwood City, CA: Addison-Wesley, 1994.

Lederman, Leon. *The God Particle: If the Universe Is the Answer, What Is the Question?* New York: Dell Publishing, 1993.

Levy, Steven. *Artificial Life: The Quest for a New Creation.* New York: Pantheon, 1992.

See Johnson, George. "After Chaos," for book review.

Lewin, Roger. *Complexity: Life At the Edge of Chaos.* New York: Macmillan Publishing, 1992. See John-son, George. "After Chaos," for book review.

Lorenz, Edward N. *The Essence of Chaos.* Seattle, WA: University of Washington Press, 1993.

"Describes how the field of knowledge he partially invented has come to be a major component in our understanding of the world around us.

Mandelbrot, Benoit B. *The Fractal Geometry of Nature.* New York: W.H. Freeman & Co., 1983.

Mazarr, Michael J. *The Revolution in Military Affairs: A Framework for Defense Planning.* Carlisle Barracks, PA: U.S. Army War College Strategic Studies Institute, June 10, 1994.

Suggests a framework for defense planning built around four pillars: 1) information dominance, 2) synergy/jointness, 3) disengaged combat, and 4) the civilianization of war. Includes a discussion on Chaos theory underlying the difficulties in predicting an RMA, which is part of a larger sociopolitical transformation.

Mazlish, Bruce. The Fourth *Discontinuity: The Co-Evolution of Humans and Machines.* New Haven: Yale University Press, 1993.

See Wagar, Warren W. for book review.

Merry, Uri. *Coping With Uncertainty: Insights From the New Sciences of Chaos, Self-Organization and Complexity.* Westport, CN: Praeger, 1995.

See Darley, Vince for book review.

Meystel, Alex. *Semiotic Modeling and Situation Analysis: An Introduction.* Bala Cynwyd, PA: AdRem, Inc, 1995.

Mueller, Theodore H. *"Chaos Theory: The Mayaguez Crisis."* U.S. Army War College, Military Studies Program,

Carlisle Barracks, PA, March 15, 1990.

Williamson, Murray and Alan R. Millett. *Military Innovation in the Interwar Period.* New York: Cambridge University Press, 1996.

Nadel, Lynn and Daniel L. Stein, eds., "1992 Lectures in Complexity Systems." *Lecture Vol. V, Santa Fe Institute Studies in the Sciences of Complexity.* Reading, MA: Addison-Wesley, 1993.

Nicolis, Gregoire and Ilya Prigogine. *Exploring Complexity: An Introduction.* New York: W.H. Freeman & Co., 1989.

Ott, Edward. *Chaos in Dynamical Systems,* 2nd Ed. Cambridge: Cambridge University Press, 1993.

      "Provides an in-depth and broad look at chaos in dynamical systems."

Ott, Edward, Tim Sauer and James A. Yorke, eds. *Coping with Chaos: Analysis of Chaotic Data and the Exploration of Chaotic Systems.* New York: John Wiley & Sons, 1994.

Pagels, Heinz. *The Dreams of Reason: The Rise of the Sciences of Complexity.* New York: Bantam, 1988.

      See Lloyd, Seth, "Complexity Simplified," for book review.

Parker, D. and Ralph Stacey. " Chaos, Management and Economics: The Implications of Nonlinear Thinking." *Hobart Papers 125*, London: Institute of Economic Affairs, 1994.

Patton, Michael Quinn. *Qualitative Evaluation and Research Methods.* Newbury Park, CA: Sage Publications, 1990.

See Chapter Three: Variety in Qualitative Inquiry: Theoretical Considerations: 64-91. Contains a section on using Chaos theory as an evaluative research approach.

Peak, David and Michael Frame. *Chaos Under Control.* New York: W.H. Freeman & Co., 1994.

"Demonstrates that the concepts and methods of analyzing fractals and chaos are not remote, theoretical brain teasers, but useful tools for understanding both the expected and unexpected dimensions of everyday experience."

Peitgen, H.O., H. Jurgens and D. Saupe. *Chaos and Fractals: New Frontiers of Science.* New York: Springer-Verlag, 1992.

"Covers the central ideas and concepts of chaos and fractals, as well as many related topics, including the Mandelbrot Set, Julia Sets, Cellular Automata, L-Systems, Percolation and Strange Attractors. "

Pentland, Pat A. *Center of Gravity Analysis and Chaos Theory, or How Societies Form, Function, and Fail.* Maxwell AFB, AL: School of Advanced Airpower Studies, AY 1993-94.

Perelson, Alan S., ed. *Theoretical Immunology, Parts One and Two.* Santa Fe Institute Studies in the Sciences Of Complexity. Proceedings vols. 2 and 3. Redwood City, CA: Addison-Wesley, 1988.

Peters, Edgar E. *Fractal Market Analysis: Applying Chaos Theory to Investment and Economics.* New York: John Wiley and Sons, 1994.

Pines, David, ed. *Emerging Syntheses in Science.* Redwood City, CA: Addison-Wesley, 1988.

Perrow, Charles. *Normal Accidents: Living With High-Risk Technologies.* New York: Basic Books, 1984.

Powers, Richard. *Galatea 2.2.* New York: Farrar Straus Giroux, 1995

    See Johnson, George for book review.

Prigonine, Ilya. *From Being to Becoming.* San Francisco, CA: W.H. Freeman, 1980.

Priesmeyer, H. Richard. *Organizations and Chaos: Defining the Methods of Nonlinear Management.* New York: Quorum Books, 1992.

    "His subtitle, *Defining the Methods of Nonlinear Management,* is very bold and sadly unjustified." There is little about organizations, and a lot about analyzing data. See McKergow for book review.

Rawlins, Gregory J. E., ed. *Foundations of Genetic Algorithms.* San Mateo, CA: Morgan Kaufmann Publishers, 1991.

Resnick, Michael. *Turtles, Termites anf Traffic Jams.* Cambridge: MIT Press, 1994.

    See Jones, Terry for book review.

Rinaldi, Steven M. *Beyond the Industrial Web: Economic Synergies and Targeting Methodologies.* Maxwell AFB, AL: Air University Press, April 1995.

"First, economies are complex systems. We can employ complexity theory to understand economic infrastructures and their behaviors. Second, given the economies of complex systems, air planners must account for their dynamics when targeting them. Finally, the reductionist methodology followed in traditional economic targeting is invalid."

Rothschild, Michael. *Bionomics: Economy as an Ecosystem.* New York: Henry Holt & Co., 1992.

Ruelle, David. *Chance and Chaos.* Princeton, NJ: Princeton University Press, 1991.

Sarigul-Klijn, Martinus M. *"Application of Chaos Methods to Helicopter Vibration Reduction Using Higher Harmonic Control."* Ph.D. dissertation, Naval Postgraduate School, Monterey, California: March 1990 (DTIC AD-A226 736).

Saul, John Ralston. *Voltaire's Bastards: The Dictatorship of Reason in the West.* New York: The Free Press, 1992.

"The Canadian novelist and essayist would probably have felt more at home in the 18th century. But he would have steered the Enlightenment toward a somewhat different conclusion. The Age of Reason, in Saul's view, has brought too much certainty to today's world, from politicians who think themselves the panacea for the world's ills to a populace mesmerized by the authority of "experts.""

As he suggests, 'we must alter our civilization from one of answers to one which feels satisfaction, not anxiety, when doubt is established.'"

Schneider, James J. *The Structure of Strategic Revolution: Total War and the Roots of the Soviet Warfare State.* Novato, CA: Presidio Press, 1994.

See Chapter One.

Schroeder, Manfred. *Fractals, Chaos, Power Laws: Minutes from an Infinite Paradise.* New York: W.H. Freeman & Co., 1991.

Snyder, Jack and Robert Jervis, eds. *Coping with Complexity in the International System.* Boulder, CO: Westview Press, 1993.

A set of thirteen essays loosely based on Complexity theory, sponsored by the Institute of War and Peace Studies, Columbia University.

Stacey, Ralph. *Strategic Management and Organizational Dynamics*, 2nd ed. London: Pitman, 1996.

Stacey, Ralph and David Parker. *Chaos, Management and Economics: the Implications of Non-linear Thinking.* IEA Hobart Paper 125, 1994.

See McKergow, Mark for book review.

Stein, Daniel L, ed. *Lectures in the Sciences of Complexity.* Redwood City, CA: Addison-Wesley, 1989.

Stewart, Ian. *Does God Play Dice? The Mathematics of Chaos.* Cambridge, MA: Basil Blackwell, 1989.

A relatively non-technical exposition of Chaos theory.

Stoppard, Tom. *Arcadia.*

A play which brings to the theater a drama which is explicitly based on Chaos theory.

Tagarev, Todor, Michael Dolgov, David Nicholls, Randal C. Franklin, and Peter Axup. *Chaos in War: Is It Present and What Does It Mean?* Report to Air Command and Staff College, Maxwell AFB, AL, Academic Year 1994 Research Program, June 1994.

Taylor, William W. *"Chaotic Evolution and Nonlinear Prediction in Signal Separation Applications."* RAND Report P-7769. Santa Monica, CA: RAND, April 1994.

Tenner, Edward. *Why Things Bite Back: New Technology and the Revenge Effect.* New York: Knopf/ Fourth Estate: 1996.

"Revenge effects differ from side-effects: 'If a cancer chemotherapy treatment causes baldness, that is not a revenge effect: but if it induces another, equally lethal cancer, that is a revenge effect.'" See Segal, Howard P., and Maddox, Bruno for book reviews.

Thompson, J.M.T. and H.B. Stewart. *Nonlinear Dynamics and Chaos.* New York: John Wiley & Sons, 1986.

Van Creveld, Martin. *Command in War.* Cambridge, MA: Harvard University Press, 1985.

Van Creveld, Martin, with Steven L. Canby and Kenneth S. Brower. *Air Power and Maneuver Warfare.* Maxwell AFB,

AL: Air University Press, July 1994.

Waldrop, M. Mitchell. *Complexity: The Emerging Science at the Edge of Order and Chaos.* New York: Simon and Schuster, 1992.

> Non-technical description of advances, personalities and institutions in the study of complex adaptive systems and spontaneous self-organization. Centers on the Santa Fe Institute. See Johnson, George and McKergow, Mark for book reviews.

Watts, Barry. *Clausewitzean Friction in Future War:* McNair Paper 52. Washington DC: Institute for National Strategic Studies, National Defense University, 1996.

> "If what counts in real war is not the absolute level of friction that either side experiences but the *relative frictional advantage* of one adversary over the other, then the question of using technology to reduce friendly friction to near zero can be seen for what it is: a false issue that diverts attention from the real business of war. Even comparatively small frictional advantages can, through nonlinear feedback, have huge consequences for combat outcomes..."

Wheatley, Margaret J. *Leadership and the New Science: Learning About Organization From an Orderly Universe.* San Francisco: Berrett-Koehler Publishers, 1992.

> "Applying revolutionary discoveries in quantum physics, chaos theory, and evolutionary biology to orga-

nizing work and people, with information at the center." Metaphoric and a whiff of New Age, but important part of the literature on "learning organizations.

Weigand, Andreas S. and Neil A. Gershenfeld, eds. *Understanding the Past: A Proceedings Volume in the Santa Fe Institute Studies in the Sciences of Complexity*. Reading, MA: Addison-Wesley, 1994.

"Compares different methods for time series prediction and characterization."

Womack, Scott Ellis. *Chaos, Clausewitz and Combat: A Critical Analysis of Operational Planning in the Vietnam War*. (Master's degree thesis.) Monterey, CA: Navy Postgraduate School, December 1995.

Wylie, Rear Admiral J.C., USN. "Military Strategy: A General Theory of Power Control," in George Edward Thibault, ed. *The Art and Practice of Military Strategy*. New York: Rutgers, The State University, 1967: 196-203.

Zurek, Wojciech H., ed. *Complexity, Entropy, and the Physics of Information*. Santa Fe Institute Studies in the Sciences of Complexity. Lectures vol 8. Redwood City, CA: Addison-Wesley, 1990.

ARTICLES, REPORTS, SPEECHES, MESSAGES

Abarbanel, Henry D.I. "Nonlinearity and Chaos at Work." *Nature,* August 19, 1993: 672-673.

Adolph, Robert B. "Playing the Numbers Game." *Army Times*, June 3, 1996:62.

Andersen, David F. "Forward: Chaos In System Dynamics Models." *System Dynamics Review,* Vol. 4, Nos. 1-2, 1988: 3-13.

     The Forward to an entire special issue of *Systems Dynamics Review* devoted to Chaos theory, consisting of ten articles.

Anderson. P. W. "More is Different," *Science,* August 4, 1972.

Antonoff, Michael. "Genetic Algorithms: Software by Natural Selection," *Popular Science,* October, 1991: 70-74.

Arthur, W. Brian. "Complexity in Economic and Financial Markets," *Complexity,* Vol. 1, No. 1: 20.

Arthur, W. Brian. "Why Do Things Become More Complex." *Scientific American,* May 1993: 144.

Arthur, W. Brian. "Pandora's Marketplace," *New Scientist,* (Supplement), February 6, 1993: S6-S8.

Arthur, W. Brian. "Positive Feedbacks in the Economy," *Scientific American,* February 1990: 92-99.

Atmanspacher, Harald, Gerda Wiedenmann and Anton Amann. "Descartes Revisited." *Complexity,* Vol. 1, No. 3: 15.

Axelrod, Robert. "An Evolutionary Approach to Norms." *American Political Science Review,* December 1986: 1095-1111.

Axelrod, Robert. "A Model of the Emergence of New Political Actors." From E. Hillebrand and J. Stender, eds. *Many-Agent Simulation and Artificial Life.* IOS Press, 1994.

Bailyn, Lotte. "Patterned Chaos in Human Resources Management." *Sloan Management Review,* Winter 1993: 77-83.

We "tend to develop elaborately structured and tightly controlled systems for managing people. Bailyn suggests a new approach: patterned Chaos. As people and their needs differ, so should their work be organized in different ways."

Bak, Per, Kan Chen and Michael Creutz. "Self-Organized Criticality in the 'Game of Life.'" *Nature*, December 14, 1989: 780-782.

Bak, Per and Kan Chen. "Self-Organized Criticality." *Scientific American*, January 1991: 46-53.

"Large interactive systems naturally evolve toward a critical state in which a minor event can lead to a catastrophe. Self-organized criticality may explain the dynamics of earthquakes, economic markets and ecosystems."

Bankes, Steve and Robert J. Lempert. "Adaptive Strategies for Abating Climate Change: An Example of Policy Analysis for Complex Adaptive Systems." Santa Monica, CA: RAND, n. d. (Missing pages)

Batterman, Robert W. "Defining Chaos." *Philosophy of Science* 60, 1993: 43-66.

Beyerchen, Alan. "Clausewitz, Nonlinearity, and the Nature of War." *International Security*, Winter 1992-93: 59-90.

Beyerchen, associated with Ohio States' Mershon Center, portrays Clausewitz as a nonlinearist. But without quantum physics and chaos theory, he "had no precise and commonly accepted vocabulary with which to express his insights into nonlinear systems."

Beard, Nick. "Evolution and Computers," *New Scientist,* January 13, 1990: 68.

Beardsley, Timothy M. "Complexity Counted." *Scientific American*, August 1988 : 16-17.

"Physicists ponder a new way to measure an elusive concept."

Begley, Sharon. "Finding Order in Disorder." *Newsweek*, December 21, 1987: 55-56.

"The science of chaos reveals nature's secrets."

Begley, Sharon. "Software au Naturel." *Newsweek*, May 8, 1995: 70-71.

"Special lines of computer code mate and mutate like living organisms. Called 'genetic algorithms,' they solve problems no human can."

Bennett, Charles H., Gilles Brassard and Arthur K. Ekert. *"Quantum Cryptography."* *Scientific American*,   October 1992: 50-57.

Bennahum, David S. "The Myth of Digital Nirvana." *Educom Review*, September-October 1996.

Excellent critique of Kevin Kelly and his products: *WIRED* magazine, and *Out of Control.*

Berlinski, David. " The Soul of Man Under Physics." *Commentary*, January 1995:38-46.

Berlinski, David. "The Deniable Darwin." *Commentary*, June 1996:19-29.

Bodanis, David. "Tender Traps for the Unwary." *New Scientist,* 19 October 1996: 52.

"...why some scientific concepts can be so difficult to understand."

Brown, Thad A. *"Political Science 443a. Nonlinear Dynamic Models."* Columbia, MO: Department of Political Science, University of Missouri. No date.

Syllabus for an advanced graduate course in the applied science of Chaos for the social sciences, with case studies.

Brown, Julian. "Where Two Worlds Meet..." *New Scientist,* 18 May 1996: 26-30.

"Butterflies can cause hurricanes, according to the classical theory of chaos. But what happens when chaos encounters the quantum world...?"

Brownlee, Shannon. "Complexity Meets the Business World." *U.S. News & World Report*, September 30, 1996: 57.

"...complexity practitioners are using the science of complexity for everything from playing the stock market to increasing the efficiency of assembly plants."

Bryngelson, Joseph D. "Provocative Questions, Problematic Answers." *Complexity*, Vol. 1, No. 2: 46.

Book review of *The Origins of Order: Self-Organization and Selection in Evolution*, by Stuart A. Kauffman.

Butler, Ann B. and James Trefill. "Evolution, Neurobiology and Behavior," *Complexity*, Vol. 1, No. 4: 3.

Calude, Cristian and Cezar Campeanu. "Are Binary Codings Universal?" *Complexity*, Vol. 1, No. 5: 47.

Carey, John. "Can the Complexity Gurus Explain It All?" *Business Week*, November 6, 1995: 23-24.

Book Review of John H. Holland's *Hidden Order* and Stuart Kaufmann's *At Home in the Universe*.

Cartwright, T. J. "Planning and Chaos Theory." *Journal of the American Planning Association*, Winter 1991: 44-56.

"Suggests that the world may be both easier and more difficult to understand than we tend to believe, that noisy and untidy cities may not be as dysfunctional as we often assume, and that the need for planning that is incremental and adaptive in nature may be more urgent than we tend to think."

Casti, John L. "Bell Curves and Monkey Languages," *Complexity*, Vol. 1, No. 1: 12.

Casti, John L. "Complexity and Simplicity: In the Eye of the Beholder," *Complexity*, Vol. 1, No. 2: 2.

Casti, John L. "If D'Arcy Only Had a Computer," *Complexity*, Vol. 1, No. 3: 5.

Casti, John L. "Seeing the Light at El Farol," *Complexity*, Vol. 1, No. 5: 7.

Chaitin, G. J. "The Berry Paradox," *Complexity*, Vol. 1, No. 1: 26.

Chaitin, G. J. "A New Version of Algorithmic Information Theory," *Complexity*, Vol. 1, No. 4: 55.

Chilcote, Ronald. "Interactive War in Vietnam: Pulverizing the Core Versus Nibbling at the Edges." (Student paper-National War College), October 1996.

"This essay is divided into four parts. The first two will examine Clausewitz's concepts of non-linearity and linearity in war. The second two will use these concepts to critique US strategy in Vietnam" through an examination of the Rolling Thunder air campaign.

Cowan, George A. "The Emergence of the Santa Fe Institute: A Complex, Adaptive System," *Complexity*, Vol. 1, No. 3: 9.

Corcoran, Elizabeth. "The Edge of Chaos." *Scientific American*, October 1992: 18-22.

Corcoran, Elizabeth. "Ordering Chaos." *Scientific American*, August 1991: 96-97.

Corcoran, Elizabeth and Paul Wallich. "Coping with Math Anxiety." *Scientific American*, August 1992: 142.

Cordesman, Anthony H. "The Quadrennial Defense Review and the American Threat to the United States." *Center for Strategic and International Studies*, January 14, 1996.

"...history and the 'chaos theory' visualized in Jurassic Park provide consistent warnings against the hubris and arrogance inherent in assuming that the right strategy and the right force posture can control our future."

Covey, Stephen R. "The Strange Attractor," *Executive Excellence*, August 1994: 5-6.

"In a properly run business, although it may look chaotic, because everyone is doing his or her own thing, they all are drawn to and united by the Strange Attractor."

Crutchfield, James P., J. Doyne Farmer, Norman H. Packard and Robert S. Shaw. "Chaos." *Scientific American*, December 1986: 46-57.

Czerwinski, Thomas J. "The Third Wave: What the Tofflers Never Told You," *Strategic Forum,* Institute for National Security Studies; National Defense University, Number 72, April 1996.

Czerwinski, Thomas J. "Command and Control at the Crossroads." *Marine Corps Gazette*, October 1995: 13-15.

Czerwinski, Thomas J. "Command and Control at the Crossroads." *Parameters*, Autumn 1996: 121-132.

Darley, Vince. "Learning to Live on the Edge." *Complexity*, Vol. 1, No. 5: 34.

Book review of *Coping With Uncertainty: Insights From the New Sciences of Chaos, Self-Organization, and Complexity*, by Uri Merry.

Davies, Paul. "Physics and the Mind of God: The Templeton Prize Address" *First Things,* August/September 1995: 31-35.

Dawid, Herbert and Alexander Mehlmann. "Genetic Learning in Strategic Form." *Complexity*, Vol. 1, No. 5: 51.

Dennard, Linda F. "The New Paradigm in Science and Public Administration." *Public Administration Review*, September/October 1996: 495-499.

Review of Margaret Wheatley's *Leadership and the New Science*, George Sessions' *Deep Ecology for the 21st Century*, and Ken Wilber's *The Holographic Paradigm and Other Paradoxes.*

Dennett, Daniel C. "Hofstadter's Quest." *Complexity*, Vol. 1, No. 6: 9.

Denning, Peter J. " Genetic Algorithms," *American Scientist,* January/February 1992: 12-14.

Dewdney, A. K., "Computer Recreations." *Scientific American,* May 1985.

Discusses cellular automata and computation.

Discover. "The Calm Before the Heart Attack," *Discover*, May 1990:

"Researchers are developing a computer program that can recognize a loss of chaos in a heartbeat—a sign that a heart attack is imminent."

Ditto, William L. and Louis M. Pecora. "Mastering Chaos." *Scientific American*, August 1993: 78-84.

"It is now possible to control some systems that behave chaotically. Engineers can use stabilize lasers, electronic circuits and even the hearts of animals."

Ditlea, Steve. "Applying Complexity Theory To Business Management." *New York Times*, February 13, 1997.

"If all this sounds a bit abstract, the clients...are decidedly down to earth. They include Mansanto, Unilever and BP Exploration." (Not to mention Citicorp and Xerox.)

Durlauf, Steven N. "Remembrance of Things Past." *Complexity*, Vol. 1, No. 3: 37-38.

Book review of *Increasing Returns and Path Dependence in the Economy*, by W. Brian Arthur.

Economist. "Tilting at Chaos." *The Economist,* August 15, 1992: 70.

Economist. "A Tale of Fat Tails." *The Economist*, October 9, 1993: 14-16.

"The theory of chaos and fractals undoubtedly describes the behavior of markets; but that does not mean it is profitable."

Evans, Karen G. "Chaos as Opportunity: Grounding a Positive Vision of Management and Society in the New Physics." *Public Administration Review*, September/October 1996: 491-494.

Review of L. Douglas Kiel's *Managing Chaos and Complexity in Government*, and Dana Zohar and Ian Marshall's *The Quantum Society: Mind, Physics and a New Social Vision.*

Forrest, Stephanie. "Genetic Algorithms: Principles of Natural Selection Applied to Computation," *Science*, August 13, 1993: 872-78.

Freeman, Walter J. "The Physiology of Perception." *Scientific American*, February 1991: 78-85.

"A familiar face, a favorite smell or a friend's voice is instantly recognized. This rapid perception depends on the coordination of millions of neurons. How can such a small input stimulate so massive a response? Surprisingly, (the answer) points to chaos—hidden order in seemingly random activity that allows many neurons to switch abruptly from one task to another."

Gell-Mann, Murray. "Let's Call It Plectics," *Complexity*, Vol. 1, No. 5: 3.

Gell-Mann, Murray. "What is Complexity?" *Complexity*, Vol. 1, No. 1: 16.

A summarization of the author's *The Quark and the Jaguar.*

Gell-Mann, Murray. "Nature Conformable to Herself," *Complexity,* Vol. 1, No. 4: 9.

Gleick, James. "Stoppard: Creating Chaos in Arcadia." *Arena Stage* Program, Washington D.C., Winter, 1996.

This theater program contains a number of commentaries on the play *Arcadia* by Tom Stoppard which contains a theme which is explicitly based on Chaos theory.

Glynn, Patrick. "Quantum Leap." *The National Interest,* Spring 1995: 50-57.

Goldberger, Ary L., David R. Rigney and Bruce J. West. "Chaos and Fractals in Human Physiology," *Scientific American*: 42-49.

Goodwin, Brian. "Emergent Form: Evolving Beyond Darwinism," *Complexity,* Vol. 1, No. 5: 11.

Gould, Stephen Jay. "The Evolution of Life on the Earth." *Scientific American,* October 1994: 85-91.

Gregersen, Hal, and Lee Sailer. "Chaos Theory and Its Implications for Social Science Research," *Human Relations* Vol. 46, No. 7, 1993: 777-802.

Grossmann, Siegfried and Gottfried Mayer-Kress. "Chaos in the International Arms Race," *Nature,* February 23, 1989: 701-4.

Gutowitz, Howard. "Cellular Automata and the Sciences of Complexity (Part I), *Complexity,* Vol. 1, No. 5: 16.

Gutowitz, Howard. "Cellular Automata and the Sciences of Complexity (Part II), *Complexity,* Vol. 1, No. 6: 29.

Gutzwiller, Martin C. "Quantum Chaos." *Scientific American*, January 1992: 78-84.

Holland, John. "Genetic Algorithms." *Scientific American*, July 1992: 66-72.

Horgan, John. "A Theory of Almost Anything." *The New York Times Book Review*, October 1, 1995: 30.

A critical review of four books: *Hidden Order: How Adaptation Builds Complexity* by John M. Holland, *At Home in the Universe: The Search for Laws of Self-Organization and Complexity* by Stuart Kauffman; *Are We Alone? Philosophical Implications of the Discovery of Extraterrestrial Life,* by Paul Davies, and *Frontiers of Complexity: The Search for Order in a Chaotic World* by Peter Coveney and Roger Highfield.

Horgan, John. "Escaping in a Cloud of Ink." *Scientific American*, August 1995: 37-41.

Profile of Stephen Jay Gould.

Horgan, John. "From Complexity to Perplexity." *Scientific American,* June 1995: 104-109.

"Can science achieve a unified theory of complex systems? Even at the Santa Fe Institute, some researchers have their doubts."

Horgan, John. "Brain Storm." *Scientific American,* November 1994: 24.

Controlling chaos could help treat epilepsy.

Horgan, John. "Nonlinear Thinking ." *Scientific American,* June 1989: 26-7.

Horgan, John. "Complexifying Freud." *Scientific American,* September 1995: 28-9.

     "Psychotherapists seek inspiration in nonlinear sciences."

Jackson, E. Atlee. "No Provable Limits to Scientific Knowledge," *Complexity,* Vol. 1, No. @: 14.

     Book review of *Chaos and Nonlinear Dynamics: An Introduction for Scientists and Engineers,* by R.C. Hilborn.

Johnson, George. "After Chaos." *Wilson Quarterly,* Spring 1993: 74-76.

     A critical review of three books—*Artificial Life* by Steven Levy, *Complexity* by M. Mitchell Waldrop, and *Complexity* by Roger Lewin.

Johnson, George. "Romancing the Brain." *Complexity,* Vol. 1, No. 5: 35-36.

     Book review of *Galatea 2.2* by Richard Powers.

Johnson, Jeffrey and Philip Picton. "How to Train a Neural Network." *Complexity,* Vol. 1, No. 6: 13.

Johnson, Phillip E. "The Storyteller and the Scientist." *Books and Culture,* November/December 1996: 46-53.

     A critical review of *The End of Science: Facing the Limits of Knowledge in the Twilight of the Scientific Age* by John Horgan.

Johnson, Phillip E. "Pomo Science." *First Things,* October 1996: 46-53.

A critical review of *Climbing Mount Probable* by Richard Dawkins and *Darwin's Black Box: The Biochemical Challenge to Evolution* by Michael Behe.

Johnson, Steven. "Strange Attraction." *Lingua Franca,* March/April 1996:42-50.

"A band of literary scholars is experimenting with the new science of Chaos. Will scientists recognize the result?"

Jones, Teresa M. "A Non-Linear Interpretation of Clausewitz and Intelligence." (Student paper-National War College), October 25, 1996.

"...investigates how Clausewitz's...*On War* addresses the issue of intelligence, explores the value of understanding the importance of non-linearity in fully comprehending his thought processes, and posits some implications for the Intelligence Community."

Jones, Terry. "In Praise of Simplicity." *Complexity,* Vol. 1, No. 3: 39.

Book review of *Turtles, Termites and Traffic Jams,* by Mitchell Resnick.

Jones, Tony. "God and Scientists Reconciled." *New Scientist,* 10 August 1996: 46.

"Ponders that fertile region where religious faith meets physics."

Kakalios, James. "Fractals: More Than Just a Pretty Picture." *Complexity*, Vol. 1, No. 5: 38-39.

Book reviews of *Fractals in Science* edited by Armin Bunde and Shlomo Havlin, and *Fractals, A User's Guide the Natural Sciences* by Harold M. Hastings and George Sugihara.

Kauffman, Stuart A. and William Macready. "Technological Evolution and Adaptive Organizations." *Complexity*, Vol. 1, No. 2: 26.

Kauffman, Stuart A. "Antichaos and Adaptation." *Scientific American*, August 1991: 78-84.

"If the tentative conclusions of this biophysicist and his colleagues are correct, there is more to evolution than natural selection. He argues that the mathematical idea of antichaos—that disorder in complex systems can suddenly crystallize into order—plays a crucial role in biology."

Kedrosky, Paul. "The More You Sell, the More You Sell." *Wired,* October 1995: 133 and 188.

"Brian Arthur's theory of 'increasing returns' is revolutionizing economics. It's also why the Department of Justice stopped the Microsoft/Intuit merger."

Kelly, Kevin. "The Economics of Ideas." *Wired,* June 1996: 148/149-217/218.

"According to economist Paul Romer, the world isn't defined by scarcity and limits on growth. Instead, it's a playground of nearly unbounded opportunity, where new ideas beget new products, new markets, and new possibilities to create wealth."

Kelly, Kevin. "The End Of Science." *Wired,* June 1996:159.

Book review of *The End of Science: Facing the Limits of Knowledge in the Twilight of the Scientific Age,* by John Horgan.

Kepler, Tom. *Complexity*, Vol. 1, No. 6: 36.

Book review of *Fire in the Mind: Science, Faith and the Search for Order,* by George M. Johnson.

Kiel, L. Douglas. "Current Thinking About Chaos Theory." *Public Administration Times,* November 1, 1995: 4 and 9.

Kiel, L. Douglas. "Nonlinear Dynamical Analysis: Assessing Systems Concepts in a Government Agency." *Public Administration Review,* March/April 1993: 143-153.

Uses "elements from Chaos theory to examine the work patterns of an Oklahoma state agency."

Kleiner, Kurt. "Fanning the Wildfires." *New Scientist,* 19 October 1996: 14-15.

"Suppressing natural fires may do more harm than good. The U.S. Forest Service now wants them to burn instead."

Lane, David A. "Models and Aphorisms." *Complexity,* Vol. 1, No. 2: 9.

Lane, David and Robert Maxfield. "Strategy Under Complexity: Fostering Generative Relationships," *Long Range Planning*, April 1996: 215-231.

Larkin, John. "Chaos Theory: It's Not Just For Scholars Anymore." *Public Administration Times*, November 1, 1995: 4 and 9.

LeBaron, Blake. "Confusion and Information on Financial Chaos." *Complexity*, Vol. 1, No. 3: 35.

Book review of *Fractal Market Analysis: Applying Chaos Theory to Investment and Economics* by Edgar E. Peters.

Lewin, Roger. "The Right Connections," *New Scientist* (Supplement), February 6, 1993: S4-S5.

Lintern, Gavan. "Complexity Stimulates Theories of Cognition and Action." *Complexity*, Vol. 1, No. 6: 38.

Book review of *A Dynamic Systems Approach to the Development of Cognition and Action,* by Esther Thelen and Linda B. Smith.

Lloyd, Seth. "Complexity Simplified," *Scientific American*, May 1996: 104-08.

Review of five books—1) *Frontiers of Complexity* by Coveney and Highfield, 2) *Hidden Order* by Holland, 3) *At Home in the Universe* by Kauffman, 4) *Fire in the Mind* by George Johnson, and 5) *The Dreams of Reason* by Heinz Pagels.

Luce, Benjamin. "Power-Packed Dynamical Systems Software." *Complexity*, Vol. 1, No. 2: 47.

Review of *Dynamics: Numerical Explorations*, by James A. Yorke and Helena E. Nusse.

Maddox, Bruno. "Monsters of Our Own Making." *Book World*, September 29, 1996: 8-9.

Review of *Why Things Bite Back: Technology and the Revenge of Unintended Consequences*, by Edward Tenner.

Mann, Steven R. "Chaos Theory and Strategic Thought." *Parameters*, Autumn 1992: 54-68.

Mark, Hans. "Some Visions For Scientific Future." Goddard Space Flight Center, October 19, 1993.

Keynote address to the Goddard Conference on Mass Storage Systems and Technologies.

Matthews, Robert. "Far Out Forecasting." *New Scientist,* 12 October 1996: 37-40.

"From the most devastating natural disasters to amazing athletics records and people who live to 124— statisticians are predicting the chances of events that border on the impossible."

McCauley, Joseph L. "Complexity, Simulations, and Emergent Law." *Phalanx*, December 1996: 8-9 and 30-31.

McCloskey, Donald M. "Once Upon a Time There Was a Theory." *Scientific American*, February 1995: 25.

McCloskey, Donald M. "Computation Outstrips Analysis." *Scientific American*, July 1995: 26.

McKergow, Mark. "Complexity Science and Management: What's in it for Business?" *Long Range Planning*, Vol. 29, No. 5, 1996: 721-727.

   A critical review of six recent books on Complexity, what they say, and how they might be applied to business.

McReady, William and David H. Wolpert. "What Makes An Optimization Problem Hard?" *Complexity*, Vol. 1, No. 5: 40.

McShea, Daniel W. "A Post-modern Vision of Artificial Life." *Complexity*, Vol. 1, No. 5: 36-37.

   Book review of *The Garden in the Machine*, by Claus Emmeche.

Mitchell, Melanie. "Genetic Algorithms: An Overview." *Complexity*, Vol. 1, No. 1: 31.

Morck, Randall and Harold Morowitz. "Value and Information: A Profit Maximizing Strategy for Maxwell's Demon." *Complexity*, Vol. 1, No. 1: 31.

Morowitz, Harold. "The Emergence of Complexity." *Complexity*, Vol. 1, No. 1: 4.

Morowitz, Harold. "The Simplicity Odyssey." *Complexity*, Vol. 1, No. 2: 7.

Morowitz, Harold. "Classified Complexity." *Complexity*, Vol. 1, No. 3: 2.

Morowitz, Harold. "What's in a Name?" *Complexity*, Vol. 1, No. 4: 7.

Morowitz, Harold. "Why Complexity Theory?" *Complexity,* Vol. 1, No. 6: 7.

Morowitz, Harold. "Back to the Future." *Complexity,* Vol. 1, No. 6: 37.

> Review of *Science and the Retreat From Reason* by John Gillott and Manjit Kumar.

Moris, Claire and Ian Langford. "No Cause for Alarm." *New Scientist,* 28 September 1996: 36-39.

> "When it comes to judging risk, most people would rather trust the opinion of a friend than take the word of a scientist." (Ties into Perrow and safety engineering principles.)

Morrison, Philip. "The Fragrance of Almonds." *Scientific American,* January 1992: 147-148.

> A review of three books on fractal art.

Murray, Williamson. "Innovation: Past and Future." *Joint Force Quarterly,* Summer 1996: 51-59.

> "Finally, the services must encourage greater familiarity with nonlinear analyses...While some suggest that the military needs more engineers to encourage nonlinear thinking, they are wrong. In fact, what the services lack are biologists, mathematicians, and historians..."

Nadis, Steve. "Poetry for Chemists." *Omni,* Fall 1995: 35.

> "A Harvard professor asks his chemistry students to write poetry to help them gain a better feel for the practice of science."

Naval Research Reviews, Office of Naval Research (ONR), Vol XLV (3), 1993

Special issue devoted to Chaos applications.

Neff, Joseph and Thomas L. Carroll. "Circuits That Get Chaos in Sync." *Scientific American*, August 1993: 120-124.

Neumann, Francis X. "Nonlinear World." *Joint Forces Quarterly*, Summer 1996: 7.

Letter to the Editor: "Having read arguments on the disestablishment of a separate Air Force in the pages of JFQ and elsewhere, I find many of them, though valid, are linear and reductionist."

Newman, Richard J. "Manual 6, Hell of a Read." *U.S. News & World Report*, December 16, 1996: 34.

"The new format is matched by an interesting new message...the Marines are incorporating new sciences such as complexity and chaos theory into their doctrine."

Nicholls, David and Todor D. Tagarev. "What Does Chaos Theory Mean for Warfare?" *Airpower Journal*, Fall 1994: 48-57.

O'Brien, Larry. "Walking at the Edge of Chaos," *AI Expert*, December 1993: 13-15.

O'Hare, Michael. "Chaos Pitch." *New Scientist*, 8 June 1996, 24-28.

Chaos theory applied to the game of soccer.

Overman, Sam E. "The New Sciences of Administration: Chaos and Quantum Theory." *Public Administration Review*, September/October 1996: 487-491.

"Chaos administration will bring new order out of chaos... Quantum administration will focus...also on energy, not matter, on becoming, not being, on intentionally, not causality, and on constructing our reality..."

Overman, Sam E. and Donna T. Loraine. "Information for Control: Another Management Proverb. " *Public Administration Review*, March/April 1994: 193-196.

Suggests "that Chaos may help us understand information as an agent of change rather than as a mechanism of control."

Paisley, Ed. "Out of Chaos, Profits." *Far Eastern Economic Review*, October 7, 1993: 84.

Park, Robert L. "At the Edge of Human Knowledge." *The Washington Post Book World*, August 11, 1996: 1-2.

Book review *of The Edge of Science: Facing the Limits of Knowledge In the Twilight of the Scientific Age*, by John Horgan.

Paulos, John Allen. "A Mathematician Critiques Popular Forecasts." *The Futurist,* November 1995: 24.

"The author debunks much of the economic and political forecasts found in popular newspapers and journals. He believes further research in Chaos theory will help improve forecasts, since they are nonlinear."

Paulos, John Allen. "Random Acts of Finance: Chaos Theory, Budget Practice." *Nation,* December 11, 1995: 752-53.

See theme above.

Pearcey, Nancy. "The Biological Challenge to Evolution." *Books and Culture,* November/December 1996: 10-11.

A critical review *of Darwin's Black Box* by Michael Behe.

Piatelli-Palminari, Massimo. "Wishful Thinking?" *Nature,* 8 August 1996: 505-06.

A review of *The Logic of Failure* by Dietrich Dorner.

Pool, Robert. "Chaos Theory: How Big an Advance?" *Science,* 7 July 1989:26-28.

"This is the last in a six-part series that examines how scientists in a host of fields are using Chaos theory to study complex phenomena. The five previous pieces, which appeared between 6 January and 10 March (1989), reported on chaos studies in epidemiology, population biology, physiology, quantum physics, and meteorology. This article explores whether chaos is merely an interesting idea enjoying a faddish vogue or is it actually...a revolution in scientific thought."

Presti, Alberto Lo. "Futures Research and Complexity." *Futures,* Vol. 28, No. 10, 1996: 891-902.

"A critical analysis from the perspective of social science."

Reilly, John J. "After Darwin." *First Things*, June-July 1995: 14-15.

Richards, Diana. "Is Strategic Decision Making Chaotic?" *Behavior Science* 35, 1990: 219-232.

Robinson, Peter. "Paul Romer." *Forbes ASAP*, June 7, 1995: 67-72.

      "Cheap powerful technology and 'free' information transforms the science of economics. But just how? This young economist knows."

Rosenau, James N. "Security in a Turbulent World." *Current History,* May 1995: 193-200.

Ruthen, Russell. "Adapting to Complexity." *Scientific American*, January 1993: 130-40.

      "What do bacteria and economies have in common? In Trying to find out, a group of multidisciplinary researchers at the Santa Fe Institute hope to derive a theory that explains why all such complex adaptive systems seem to evolve toward the boundary between order and chaos. Their ideas could result in a view of evolution that encompasses living and nonliving systems."

Santa Fe Institute. *Bulletin of the Santa Fe Institute* (1987-present). Published two or three times a year. Contains extended interviews, and summaries of workshops and meetings.

Saperstein, Alvin M. "War and Chaos." *American Scientist,* November-December 1995: 548-557.

"Complexity theory may be useful in modeling how real-world situations get out of control."

Scheeline, Alexander and Yeou-Teh Liu. "Chaos Limited." *Complexity*, Vol. 1, No. 1: 48.

Review of *Chaos Demonstrations, V. 2.0 and Chaos Data Analyzer, V. 1.0,* by J.C. Sprott and G. Rowlands.

Scheeline, Alexander and Nicholas Weber. "Send in the Clones." *Complexity*, Vol. 1, No. 2: 48.

Review of *Modelmaker, V. 2 ,* by SB Technology.

Schwartz, John. "Taking Advantage of Chaos to Find Stability and Maintain Control." *Washington Post*, July 4, 1994: A3.

Segal, Howard P. "Expecting the Unexpected." *Nature,* 8 August 1996: 504-05.

A review of *Why Things Bite Back* by Edward Tenner.

Segel, Lee A. "Grappling with Complexity." *Complexity,* Vol. 1, No. 2: 18.

Shepard, Harvey. "Why the World is Simple." *Complexity*, Vol. 1, No. 1: 46.

Review of *The Collapse of Chaos*, by Jack Cohen and Ian Stewart.

Shinbrot, Troy, Celso Grebogi, Edward Ott and James A. Yorke. "Using Small Perturbations to Control Chaos." *Nature,* June 3, 1993: 415.

Sigmund, Karl. "Darwin's 'Circles of Complexity': Assembling Ecological Communities." *Complexity*, Vol. 1, No. 1: 40.

Silk, Joseph I. "Road to Nowhere." *Scientific American*, July 1995: 93-94.

   A review of *The Physics of Immortality* by Frank J. Tippler.

Smith, Douglas. "How to Generate Chaos at Home." *Scientific American,* January 1992: 144-146.

   Building an electronic circuit which when subjected to certain voltages, produces a signal that is chaotic.

Smith, R. David. *"The Inapplicability Principal: What Chaos Means for Social Science."* Behavioral Science, Vol. 40, 1995: 22.

Sole, Ricard V. "On Macroevolution, Extinctions and Critical Phenomena." *Complexity*, Vol. 1, No. 6: 40.

Sole, Ricard V., Sussanna C. Manrubia, Bartolo Lique, Jordi Delgado and Jordi Bascompte. "Phase Transitions and Complex Systems." *Complexity*, Vol. 1, No. 4: 13.

Somogyi, Roland and Carol Ann Sniegoski. "Modeling the Complexity of Genetic Networks." *Complexity,* Vol. 1, No. 6: 45.

Stacey, Ralph. "Emerging Strategies for a Chaotic Environment," *Long Range Planning*, April 1996: 182-189.

Stacey, Ralph. "The Science of Complexity: An Alternative Perspective for Strategic Change Processes." *The Strategic Management Journal*, August, 1995.

Sterman, John D. "Deterministic Chaos in Models of Human Behavior: Methodological Issues and Experimental Results." *Systems Dynamics Review* 4 (1-2), 1988: 148-178.

Stewart, Ian. "Christmas In the House of Chaos." *Scientific American,* December 1992: 144-146.

A tale of holiday fractal art and Christmas tree decorating.

Svozil, K. "How Real are Virtual Realities, How Virtual is Reality?—Constructive Re-interpretation of Physical Undecidability." *Complexity*, Vol. 1, No. 4: 43.

Tagg, Randall. "A Field Guide to Chaos." *Complexity*, Vol. 1, No. 1: 45.

A review of *The Nature of Chaos,* edited by Tom Mullin.

Taylor, William C. "Control in an Age of Chaos." *Harvard Business Review*, November-December 1994: 64-76.

"The New Economy demands new models of organization." A review and synthesis of three works; *Out of Control: The Rise of Neo-Biological Civilization* by Kevin Kelly, *Built to Last: Successful Habits of Visionary Companies* by Jasmes C. Collins and Jerry I. Porras, and

Showstopper! The Breakneck Race to Create Windows NT and the Next Generation at Microsoft by G. Pascal Zachary.

Tepper, August. "Controlling Technology by Shaping Visions," *Policy Sciences*, Vol. 29, No. 1, 1996: 29-44.

Excellent survey of the use and abuse of metaphors.

Tempestelli, Mark. "The Network Force." *Proceedings*, June 1996:42-46.

Testa, Bernard and Lemont B. Kier. "Complex Systems in Drug Research." *Complexity*, Vol. 1, No. 4: 29.

Tsonis, A.A. "Dynamical Systems as Models for Physical Processes." *Complexity*, Vol. 1, No. 5: 23.

Varian, Hal R. "The Information Economy." *Scientific American,* September 1995: 200-201.

Vincent, Gary A. "A New Approach to Command and Control: The Cybernetic Design." *Airpower Journal*, Summer 1993: 24-38.

Argues for the elimination of "intermediate control units" which would thus minimize friction. The aim is to exploit advances in technology in order to overcome the challenges of survivability and speed.

Voohees, Burton. "Godel's Theorem and the Possibility of Thinking Machines." *Complexity*, Vol. 1, No. 3: 30.

Watts, Barry D. "Friction in the Gulf War." *Naval War College Review*, Autumn 1995, 93-108.

A review of Gordon and Trainor's *The Generals' War* wherein Watts grounds Clausewitzean friction in a nonlinear science context.

Wayner, Peter. "Genetic Algorithms," *Byte,* January 1991: 361-68.

Weiss, Gary. "Chaos Hits Wall Street-The Theory, That Is." *Business Week*, November 2, 1992: 138-140.

"An arcane market system is making waves."

Wheatley, Margaret J. "A Quantum Vision: Chaotic Organization Must Replace the Newtonian Bureaucracy." *Barron's*, March 22, 1993: 12.

Wheatley, Margaret J. "Can the U.S. Army Become a Learning Organization? *Journal of Quality and Participation,* XXX: XX.

Wiley & Sons. *Complexity.* Published bimonthly, starting July 1995, this publication offers recent research results, educational overviews, tutorials and book/article reviews.

Wolfram, Stephen. "Computer Software in Science and Mathematics." *Scientific American*, September 1984.

Contains a basic description of cellular automata.

Yam, Philip. "Chaotic Chaos." *Scientific American*, March 1994: 16.

Zhang, Shuguang and Martin Egli. "A Possible Pathway for Generating Complex Biological Molecules." *Complexity*, Vol. 1, No. 1: 49.

Zurich, Wojciech Hubert. "The Many Faces of Information." *Complexity*, Vol. 1, No. 2: 64.

BRIEFS

Ilachinski, Andy. "Land Warfare as a Complex Adaptive System." Center for Naval Analysis, 1995.